FUNDAMENTALS OF METEOROLOGY

Edited by
Judith Rosales

Delve Publishing

Fundamentals of Meteorology

Edited by Judith Rosales

Edition 2018

Delve Publishing

Published by
Delve Publishing LLC
708 3rd Avenue, 6th Floor
New York
NY 10801
United States of America
www.delvepublishing.com

ISBN: 978-1-68095-738-9 (Hardback)

Printed in United States of America

Library of Congress Control Number: 2016961908

About the Editor

Judith Rosales

Prof. Judith is an Ecologist and Environmental Specialist with more than 30 years of academic and consulting experience. She holds a PhD in geography from the University of Birmingham UK and wrote the PhD and Master Academic Programs of Environmental Sciences for the Universidad de Guayana, Venezuela.

Table of Contents

2 SURFACE WEATHER ANALYSIS

3 WEATHER OBSERVATION AND PREDICTION OF WEATHER

4 WEATHER MAP

5 METEOROLOGY

6 ATMOSPHERIC CIRCULATION

7 METEOROLOGICAL PHENOMENA

List of Abbreviations

AGCM	Atmospheric General Circulation Model
CP	Central Pacific
CAIR	Clean Air Interstate Rule
EP	Eastern Pacific
ENSO	'El Niño–Southern Oscillation'
EPA	Environmental Protection Agency's
EQSOI	Equatorial Southern Oscillation Index
ECMWF	European Centre for Medium-Range Weather Forecasts
FGD	Flue-Gas Desulfurization
GCM	General Circulation Model
HM	Hydrographic and Meteorological
JNWPU	Joint Numerical Weather Prediction Unit
MJO	Madden–Julian Oscillation
MSLP	Mean Sea Level Pressure
MCS	Mesoscale Convective System
MS	Meteorological Service
MOGREPS	Met Office Global and Regional Ensemble Prediction System
MOS	Model Output Statistics,
MFM	Movable Fine-mesh
NAPAP	National Acidic Precipitation Assessment Program
NCAR	National Center for Atmospheric Research

NOAA	National Oceanic and Atmospheric Administration
NWS	National Weather Service
NWP	Numerical Weather prediction
SO	Sulfur Dioxide
TWC	The Weather Channel
TNI	Trans-Niño index
UV	Ultraviolet Radiation
WRF	Weather Research and Forecasting Model
WMO	World Meteorological Organization

Preface

Meteorology is the scientific study of the atmosphere that focuses on weather processes and forecasting. Meteorological phenomena are observable weather events which illuminate and are explained by the science of meteorology. It is a study that dates back to ancient times, when ancient civilizations made observations and kept records of weather conditions, both for agricultural purposes and out of a general curiosity about the world around them. Over the centuries, the atmosphere has been studied for a variety of reasons, including agricultural knowledge, military defense and planning, and developing better warnings for severe weather systems like tornadoes and hurricanes. Technological advances, such as the development of scientific computing and an increase in the total number of meteorological observations being taken daily across the globe, have allowed for better forecasts (or at least the meteorological community likes to think they are better forecasts) and a much better overall understanding of our atmosphere. Meteorological phenomena are observable weather events that are explained by the science of meteorology. Meteorological phenomena are described and quantified by the variables of Earth's atmosphere: temperature, air pressure, water vapor, mass flow, and the variations and interactions of those variables, and how they change over time. Different spatial scales are used to describe and predict weather on local, regional, and global levels. There are many specialized branches, or subfields, of meteorology, each defined by both the spatial scale of the phenomena they study and by the other branches of science they involve. In terms of spatial scale, all meteorological study is conducted on either the micro-, meso-, synoptic, or global scale. Microscale meteorology deals with any problems that are on the order of 1 kilometer in size or smaller, things like individual clouds or heat transfers. Mesoscale meterology includes phenomena in the size range from the upper end of the microscale to several hundred kilometers. Lake effect snowstorms, severe thunderstorms, and mesoscale convective complexes.

Content Coverage

Chapter One is intended to present the fundamentals of atmosphere. The atmosphere helps protect living organisms from genetic damage by solar ultraviolet radiation, solar wind and cosmic rays.

Chapter Two focuses on surface weather analysis, which is a special type of weather map that provides a view of weather elements over a geographical area at a specified time based on information from ground-based weather stations.

Chapter Three explores the methods of the acquisition of weather data used for research and planning. Weather forecasting is the application of science and technology to predict the state of the atmosphere for a given location. Human beings have attempted to predict the weather informally for millennia, and formally since the nineteenth century.

Chapters Four and *Five* highlight the weather maps and meteorology perspectives, respectively. Weather maps have been in use since the mid-19th century and are used for research and weather forecasting purposes. Meteorological phenomena, however, are observable weather events that are explained by the science of meteorology.

Chapter Six focuses on the Earth's atmospheric circulation varies from year to year, but the large scale structure of its circulation remains fairly constant.

Chapter Seven explores the meteorological phenomenon, which is a weather event that can be explained by the principles of meteorology. Such events include acid rain, air mass, etc.

Chapter Eight presents an overview of microscale meteorology, the study of weather systems smaller than synoptic scale systems but larger than microscale and storm-scale cumulus systems. Horizontal dimensions generally range from around 5 kilometers to several hundred kilometers.

CHAPTER 1

An Overview of Atmospheric Fundamentals

Chapter Contents

The highly interdisciplinary Atmospheric Sciences specialization is organized by both the Department of Earth, Ocean and Atmospheric Science and the Department of Geography. Additional professors in our program are from the departments of Chemistry, Math, Soil Science, and Engineering. The 13 professors in the Atmospheric Sciences program conduct research and instruction in the broad fields of atmospheric science, physical & biological oceanography, climatology, glaciology, biometeorology, applied meteorology, and clean-energy engineering.

INTRODUCTION

An atmosphere is a layer of gases surrounding a planet or other material body, that is held in place by the gravity of that body. An atmosphere is more likely to be retained if the gravity it is subject to is high and the temperature of the atmosphere is low.

The atmosphere of Earth is mostly composed of nitrogen (about 78%), oxygen (about 21%), argon (about 0.9%) with carbon dioxide and other gases in trace amounts. Oxygen is used by most organisms for respiration, nitrogen is fixed by bacteria and lightning to produce ammonia used in the construction of nucleotides and amino acids and carbon dioxide is used by plants, algae and cyanobacteria for photosynthesis. The atmosphere helps protect living organisms from genetic damage by solar ultraviolet radiation, solar wind and cosmic rays. Its current composition is the product of billions of years of biochemical modification of the paleoatmosphere by living organisms.

The term stellar atmosphere describes the outer region of a star, and typically includes the portion starting from the opaque photosphere outwards. Stars with sufficiently low temperatures may form compound molecules in their outer atmosphere.

Mars's thin atmosphere

ATMOSPHERIC PRESSURE

Atmospheric pressure, sometimes also called barometric pressure, is the pressure exerted by the weight of air in the atmosphere of Earth (or that of another planet). In most circumstances atmospheric pressure is closely approximated by the hydrostatic pressure caused by the weight of air above the measurement point. Low-pressure areas have less atmospheric mass above their location, whereas high-pressure areas have more atmospheric mass above their location. Likewise, as elevation increases, there is less overlying atmospheric mass, so that atmospheric pressure decreases with increasing elevation. On average, a column of air one square centimetre [cm2] (0.16 sq in) in cross-section, measured from sea level to the top of the atmosphere, has a mass of about 1.03 kilograms (2.3 lb) and weight of about 10.1 newtons (2.3 lbf). That force (across one square centimeter) is a pressure of 10.1 N/cm2 or 101,000 N/m2. A column 1 square inch (6.5 cm2) in cross-section would have a weight of about 14.7 lb (6.7 kg) or about 65.4 N.

Standard Atmospheric

The standard atmosphere (symbol: atm) is a unit of pressure defined as 101325 Pa (1.01325 bar), equivalent to 760 mmHg (torr), 29.92 inHg and 14.696 psi.

Mean Sea Level Pressure

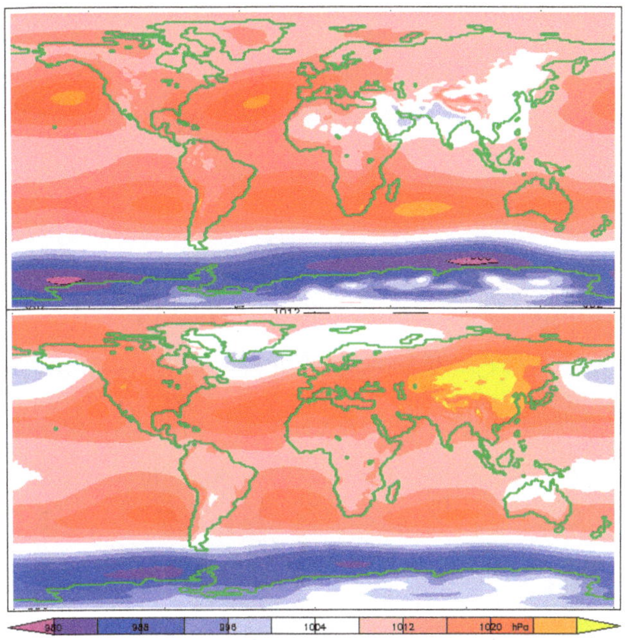

15 year average mean sea level pressure for June, July, and August (top) and December, January, and February (bottom). ERA-15 re-analysis.

Kollsman-type barometric aircraft altimeter (as used in North America) displaying an altitude of 80 ft (24 m).

The mean sea level pressure (MSLP) is the atmospheric pressure at sea level. This is the atmospheric pressure normally given in weather reports on radio, television, and newspapers or on the Internet. When barometers in the home are set to match the local weather reports, they measure pressure adjusted to sea level, not the actual local atmospheric pressure.

The *altimeter setting* in aviation, is an atmospheric pressure adjustment.

Average *sea-level pressure* is 1013.25 mbar (101.325 kPa; 29.921 inHg; 760.00 mmHg). In aviation weather reports (METAR), QNH is transmitted around the world in millibars or hectopascals (1 hectopascal = 1 millibar), except in the United States, Canada, and Colombia where it is reported in inches (to two decimal places) of mercury. The United States and Canada also report *sea level pressure* SLP, which is adjusted to sea level by a different method, in the remarks section, not in the internationally transmitted part of the code, in hectopascals or millibars. However, in Canada's public weather reports, sea level pressure is instead reported in kilopascals.

In the US weather code remarks, three digits are all that are transmitted; decimal points and the one or two most significant digits are omitted: 1013.2 mbar (101.32 kPa) is transmitted as 132; 1000.0 mbar (100.00 kPa) is transmitted as 000; 998.7 mbar is transmitted as 987; etc. The highest *sea-level pressure* on Earth occurs in Siberia, where the Siberian High often attains a *sea-level pressure* above 1050 mbar (105 kPa; 31 inHg), with record highs close to 1085 mbar (108.5 kPa; 32.0 inHg). The lowest measurable *sea-level pressure* is found at the centers of tropical cyclones and tornadoes, with a record low of 870 mbar (87 kPa; 26 inHg) (see Atmospheric pressure records).

Altitude Variation

A very local storm above Snæfellsjökull, showing clouds formed on the mountain by orographic lift

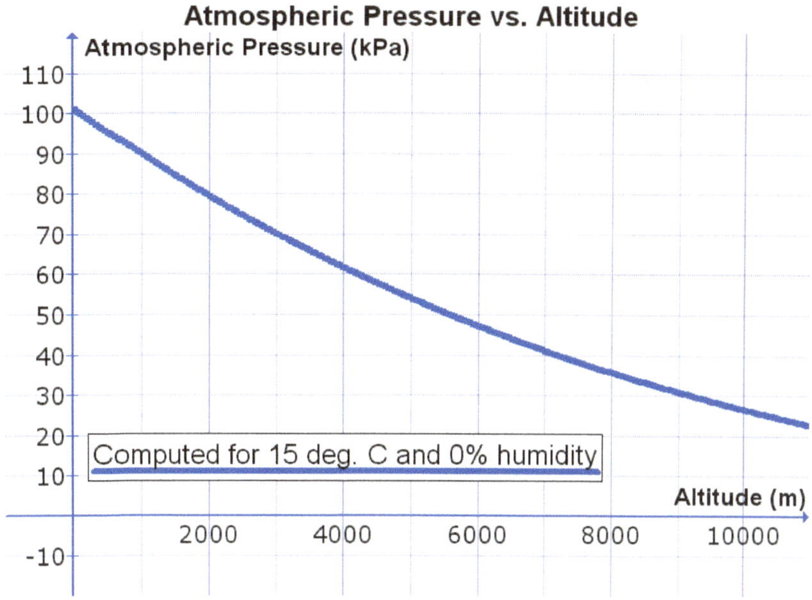

Variation in atmospheric pressure with altitude, computed for 15 °C and 0% relative humidity.

This plastic bottle was sealed at approximately 14,000 feet (4,300 m) altitude, and was crushed by the increase in atmospheric pressure —at 9,000 feet (2,700 m) and 1,000 feet (300 m)— as it was brought down towards sea level.

Pressure varies smoothly from the Earth's surface to the top of the mesosphere. Although the pressure changes with the weather, NASA has averaged the conditions for all parts of the earth year-round. As altitude increases, atmospheric pressure decreases. One can calculate the atmospheric pressure at a given altitude. Temperature and humidity also affect the atmospheric pressure, and it is necessary to know these to compute an accurate figure. The graph at right was developed for a temperature of 15 °C and a relative humidity of 0%.

At low altitudes above the sea level, the pressure decreases by about 1.2 kPa for every 100 meters. For higher altitudes within the troposphere, the following equation (the barometric formula) relates atmospheric pressure p to altitude h

$$p = p_0 \cdot \left(1 - \frac{L \cdot h}{T_0}\right)^{\frac{g \cdot M}{R_0 \cdot L}} \approx p_0 \cdot \left(1 - \frac{g \cdot h}{c_p \cdot T_0}\right)^{\frac{c_p \cdot M}{R_0}},$$

$$p \approx p_0 \cdot \exp\left(-\frac{g \cdot M \cdot h}{R_0 \cdot T_0}\right)$$

where the constant parameters are as described below:

Parameter	Description	Value
p_0	sea level standard atmospheric pressure	101325 Pa
L	temperature lapse rate, = g/c_p for dry air	0.0065 K/m
c_p	constant pressure specific heat	~ 1007 J/(kg•K)
T_0	sea level standard temperature	288.15 K
g	Earth-surface gravitational acceleration	9.80665 m/s^2
M	molar mass of dry air	0.0289644 kg/mol
R_0	universal gas constant	8.31447 J/(mol•K)

Local Variation

Hurricane Wilma on 19 October 2005; 882 hPa (12.79 psi) in the storm's eye

Atmospheric pressure varies widely on Earth, and these changes are important in studying weather and climate. See pressure system for the effects of air pressure variations on weather.

Atmospheric pressure shows a diurnal or semidiurnal (twice-daily) cycle caused by global atmospheric tides. This effect is strongest in tropical zones, with an amplitude of a few millibars, and almost zero in polar areas. These variations have two superimposed cycles, a circadian (24 h) cycle and semi-circadian (12 h) cycle.

Records

The highest adjusted-to-sea level barometric pressure ever recorded on Earth (above 750 meters) was 1,085.7 hectopascals (32.06 inHg) measured in Tosontsengel, Mongolia on 19 December 2001. The highest adjusted-to-sea level barometric pressure ever recorded (below 750 meters) was at Agata, Evenkiyskiy, Russia 66°53'N, 93°28'E, elevation: 261 m (856.3 ft) on 31 December 1968 of 1,083.8 hectopascals (32.00 inHg). The discrimination is due to the problematic assumptions (assuming a standard lapse rate) associated with reduction of sea level from high elevations.

The Dead Sea, the lowest place on Earth at 425 metres (1400 feet) below sea level, has a correspondingly high typical atmospheric pressure of 1065 hPa.

The lowest non-tornadic atmospheric pressure ever measured was 0.858 atm (25.69 inHg), 870 hPa, set on 12 October 1979, during Typhoon Tip in the western Pacific Ocean. The measurement was based on an instrumental observation made from a reconnaissance aircraft.

Measurement Based on Depth of Water

One atmosphere (101 kPa or 14.7 psi) is the pressure caused by the weight of a column of fresh water of approximately 10.3 m (33.8 ft). Thus, a diver 10.3 m underwater experiences a pressure of about 2 atmospheres (1 atm of air plus 1 atm of water). Conversely, 10.3 m is the maximum height to which water can be raised using suction under standard atmospheric conditions.

Low pressures such as natural gas lines are sometimes specified in inches of water, typically written as *w.c.* (water column) or *w.g.* (inches water gauge). A typical gas-using residential appliance in the US is rated for a maximum of 14 w.c., which is approximately 35 hPa. Similar metric units with a wide variety of names and notation based on millimetres, centimetres or metres are now less commonly used.

Boiling Point of Water

Pure water boils at 100 °C (212 °F) at earth's standard atmospheric pressure. The boiling point is the temperature at which the vapor pressure is equal to the atmospheric pressure around the water. Because of this, the boiling point of water is lower at lower pressure and higher at higher pressure. Cooking at high elevations, therefore, requires adjustments to recipes. A rough approximation of elevation can be obtained by measuring the temperature at which water boils; in the mid-19th century, this method was used by explorers.

Boiling water

Measurement and Maps

An important application of the knowledge that atmospheric pressure varies directly with altitude was in determining the height of hills and mountains thanks to the availability of reliable pressure measurement devices. While in 1774 Maskelyne was confirming Newton's theory of gravitation at and on Schiehallion in Scotland (using plumb bob deviation to show the effect of "gravity") and accurately measure elevation, William Roy using barometric pressure was able to confirm his height determinations, the agreement being to within one meter (3.28 feet). This was then a useful tool for survey work and map making and long has continued to be useful. It was part of the "application of science" which gave practical people the insight that applied science could easily and relatively cheaply be "useful".

ATMOSPHERIC ESCAPE

Atmospheric escape is the loss of planetary atmospheric gases to outer space. A number of different mechanisms can be responsible for atmospheric escape, operating at different time scales; the most prominent is Jeans Escape, named after British astronomer Sir James Jeans, who described the process of atmospheric loss to the molecular kinetic energy..

Thermal Escape Mechanisms

One classical thermal escape mechanism is **Jeans escape**. In a quantity of gas, the average velocity of a molecule is determined by temperature, but the velocities of individual molecules change as they collide with one another, gaining and losing kinetic energy. The variation in kinetic energy among the molecules is described by the Maxwell distribution. The kinetic energy and mass of a molecule determine its velocity by $E_{kin} = \frac{1}{2}mv^2$.

Individual molecules in the high tail of the distribution may reach escape velocity, at a level in the atmosphere where the mean free path is comparable to the scale height, and leave the atmosphere.

The more massive the molecule of a gas is, the lower the average velocity of molecules of that gas at a given temperature, and the less likely it is that any of them reach escape velocity.

This is why hydrogen escapes from an atmosphere more easily than carbon dioxide. Also, if the planet has a higher mass, the escape velocity is greater, and fewer particles will escape. This is why the gas giant planets still retain significant amounts of hydrogen and helium, which have largely escaped from Earth's atmosphere. The distance a planet orbits from a star also plays a part; a close planet has a hotter atmosphere, with a range of velocities shifted into the higher end of the distribution, hence, a greater likelihood of escape. A distant body has a cooler atmosphere, with a range of lower velocities, and less chance of escape. This helps Titan, which is small compared to Earth but further from the Sun, retain its atmosphere.

An atmosphere with a high enough pressure and temperature can undergo a different escape mechanism - "hydrodynamic escape". In this situation the atmosphere simply flows off like a wind into space, due to pressure gradients initiated by thermal energy deposition. Here it is possible to lose heavier molecules that would not normally be lost. Hydrodynamic escape has been observed for exoplanets close-to their host star, including several hot Jupiters (HD 209458b, HD 189733b) and a hot Neptune (GJ 436b).

Significance of Solar Winds

The relative importance of each loss process is a function of planet mass, its atmosphere composition, and its distance from its sun. A common erroneous belief is that the primary non-thermal escape mechanism is atmospheric stripping by a solar wind in the absence of a magnetosphere. Excess kinetic energy from solar winds can impart sufficient energy to the atmospheric particles to allow them to reach escape velocity, causing atmospheric escape. The solar wind, composed of ions, is deflected by magnetic fields because the charged particles within the wind flow along magnetic field lines. The presence of a magnetic field thus deflects solar winds, preventing the loss of atmosphere. On Earth, for instance, the interaction between the solar wind and earth's magnetic field deflects the solar wind about the planet, with near total deflection at a distance of 10 Earth radii. This region of deflection is called a bow shock.

Depending on planet size and atmospheric composition, however, a lack of magnetic field does not determine the fate of a planet's atmosphere. Venus, for instance, has no powerful magnetic field. Its close proximity to the Sun also increases the speed and number of particles, and would presumably cause the atmosphere to be stripped almost entirely, much like that of Mars. Despite this, the atmosphere of Venus is two orders of magnitudes denser than Earth's. Recent models indicate that stripping by solar wind accounts for less than 1/3 of total non-thermal loss processes.

While Venus and Mars have no magnetosphere to protect the atmosphere from solar winds, photoionizing radiation (sunlight) and the interaction of the solar wind with the atmosphere of the planets causes ionization of the uppermost part of the atmosphere. This ionized region in turn induces magnetic moments that deflect solar winds much like a magnetic field. This limits solar-wind effects to the uppermost part of the atmosphere, roughly 1.2–1.5 planetary radii away from the planet, or an order of magnitude closer to the surface than Earth's magnetic field creates. Beyond this region, called a bow shock, the solar wind is slowed to subsonic velocities. Nearer to the surface, solar-wind dynamic pressure reaches a balance with the pressure from the ionosphere, in a region called the ionopause. This interaction typically prevents solar-wind stripping from being the dominant loss process of the atmosphere.

Comparison of Non-Thermal Loss Processes Based on Planet and Particle Mass

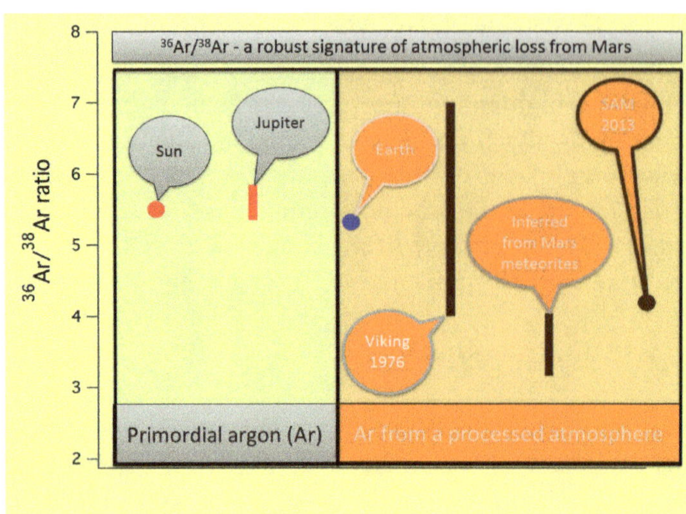

Argon Isotope Ratios are a signature of atmospheric escape on Mars (Curiosity rover) (April, 2013)

The dominant non-thermal loss processes depends on the planetary body. The relative significance of each process depends on planetary mass, atmospheric composition, and distance from the sun. The dominant non-thermal loss processes for Venus and Mars, two terrestrial planets neither with magnetic fields, are dissimilar. The dominant non-thermal loss process on Mars is from solar winds, as the atmosphere is not dense enough to shield itself from the winds during peak solar activity. Venus is somewhat shielded from solar winds because of its denser atmosphere and as a result, solar pick-up is not its dominant non-thermal loss process. Smaller bodies without magnetic fields are more likely to suffer from solar winds, as the planet is too small to have sufficient gravity to produce a dense enough atmosphere and stop solar wind pick-up.

The dominant loss process for Venus' atmosphere is through electric force field acceleration. As electrons are less massive than other particles, they are more likely to escape from the top of Venus's ionosphere. As a result, a minor net positive charge develops. That net positive charge, in turn, creates an electric field that can accelerate other positive charges out of the atmosphere. As a result, H^+ ions are accelerated beyond escape velocity. Other important loss processes on Venus are photochemical reactions driven by Venus's proximity to the Sun. Photochemical reactions rely on the splitting of molecules into constituent atoms, often with a significant portion of the kinetic energy carried off in the less massive particle with sufficiently high kinetic energy to escape. Oxygen, relative to hydrogen, is not of sufficiently low mass to escape through this mechanism.

Phenomena of Non-Thermal Loss Processes on Moons with Atmospheres

Several natural satellites in the Solar System have atmospheres and are subject to atmospheric loss processes. They typically have no magnetic fields of their own, but orbit planets with powerful magnetic fields. Many of these moons lie within the magnetic fields generated by the planets and are less likely to undergo sputtering and pick-up. The shape of the bow shock, however, allows for some moons, such as Titan, to pass through the bow shock when their orbits take them between the Sun and their primary. Titan spends roughly half of its transit time outside of the bow-shock and being subjected to unimpeded solar winds. The kinetic energy gained from pick-up and sputtering associated with the solar winds increases thermal escape throughout the transit of Titan, causing neutral hydrogen to escape. The escaped hydrogen maintains an orbit following in the wake of Titan, creating a neutral hydrogen torus around Saturn. Io, in its transit around Jupiter, encounters a plasma cloud. Interaction with the plasma cloud induces sputtering, kicking off sodium particles. The interaction produces a stationary banana-shaped charged sodium cloud along a part of the orbit of Io.

Impact Erosion

The impact of a large meteoroid can lead to the loss of atmosphere. If a collision is energetic enough, it is possible for ejecta, including atmospheric molecules, to reach escape velocity. Just one impact such as the Chicxulub event does not lead to a significant loss, but the terrestrial planets went through enough impacts when they were forming for this to matter.

Sequestration

Sequestration is not a form of escape from the planet, but a loss of molecules from the atmosphere and into the planet. It occurs on Earth when water vapor condenses to form rain or glacial ice. It also occurs on Earth when carbon dioxide is sequestered in sediments, or cycled through the oceans. The dry ice caps on Mars are also an example of sequestration.

One mechanism for sequestration is chemical; for example, most of the carbon dioxide of the Earth's original atmosphere has been chemically sequestered into carbonate rock. Very likely a similar process has occurred on Mars. Oxygen can be sequestered by oxidation of rocks; for example, by increasing the oxidation states of ferric rocks from Fe^{2+} to Fe^{3+}. Gases can also be sequestered by adsorption, where fine particles in the regolith capture gas which adheres to the surface particles.

Dominant Atmospheric Escape and Loss Processes on Earth

Earth is too large to lose a significant proportion of its atmosphere through Jeans escape. The current rate of loss is about three kilograms (3 kg) of hydrogen and 50 grams (50 g) of helium per second. The exosphere is the high-altitude region where atmospheric density is sparse and Jeans escape occurs. Jeans escape calculations assuming an exosphere temperature of 1,800 K show that to deplete O^+ ions by a factor of *e* (2.718...) would take nearly a billion years. 1,800 K is higher than the actual observed exosphere temperature; at the actual average exosphere temperature, depletion of O^+ ions would not occur even over a trillion years. Furthermore, most oxygen on Earth is bound as O_2, which is too massive to escape Earth by Jeans escape.

Earth's magnetic field protects it from solar winds and prevents escape of ions, except near the magnetic poles where charged particles stream towards the earth along magnetic field lines. The gravitational attraction of Earth's mass prevents other non-thermal loss processes from appreciably depleting the atmosphere. Yet Earth's atmosphere is two orders of magnitude less dense than that of Venus at the surface. Because of the temperature regime of Earth, CO_2 and H_2O are sequestered in the hydrosphere and lithosphere. H_2O vapor is sequestered as liquid H_2O in oceans, greatly decreasing the atmospheric density. With liquid water running over the surface of Earth, CO_2 can be drawn down from the atmosphere and sequestered in sedimentary rocks. Some estimates indicate that nearly all carbon on Earth is contained in sedimentary rocks, with the atmospheric portion being approximately 1/250,000 of Earth's CO_2reservoir. If both of the reservoirs were released to the atmosphere, Earth's atmosphere would be even denser than Venus's atmosphere. Therefore, the dominant "loss" mechanism of Earth's atmosphere is not escape to space, but sequestration.

Terrain

Atmospheres have dramatic effects on the surfaces of rocky bodies. Objects that have no atmosphere, or that have only an exosphere, have terrain that is covered in craters.

Without an atmosphere, the planet has no protection from meteors, and all of them collide with the surface and create craters.

A rocky body with a thick atmosphere does not have significant craters on its surface. The friction generated when a meteor enters an atmosphere causes the vast majority of it to burn up before hitting the surface. When meteors do impact, the effects are often erased by the action of wind. As a result, craters are rare on objects with atmospheres.

All objects with atmospheres have wind and weather. Wind erosion is a significant factor in shaping the terrain of rocky planets with atmospheres, and over time can erase the effects of both craters and volcanoes. In addition, since liquids can not exist without pressure, an atmosphere allows liquid to be present at the surface, resulting in lakes, rivers and oceans. Earth and Titan are known to have liquids at their surface and terrain on the planet suggests that Mars had liquid on its surface in the past.

Composition

Earth's atmospheric gases scatter blue light more than other wavelengths, giving Earth a blue halo when seen from space

Initial atmospheric composition is generally related to the chemistry and temperature of the local solar nebula during planetary formation and the subsequent escape of interior gases. The original atmospheres started with the radially local rotating gases that collapsed to the spaced rings that formed the planets. They were then modified over time by various complex factors, resulting in quite different outcomes.

The atmospheres of the planets Venus and Mars are primarily composed of carbon dioxide, with small quantities of nitrogen, argon, oxygen and traces of other gases.

The atmospheric composition on Earth is largely governed by the by-products of the life that it sustains. Dry air from Earth's atmosphere contains 78.08% nitrogen, 20.95% oxygen, 0.93% argon, 0.04% carbon dioxide, and traces of hydrogen, helium, and other "noble" gases (by volume), but generally a variable amount of water vapour is also present, on average about 1% at sea level.

The low temperatures and higher gravity of the Solar System's giant planets—Jupiter, Saturn, Uranus and Neptune—allow them more readily to retain gases with low molecular masses. These planets have hydrogen–helium atmospheres, with trace amounts of more complex compounds.

Two satellites of the outer planets possess significant atmospheres. Titan, a moon of Saturn, and Triton, a moon of Neptune, have atmospheres mainly of nitrogen. When in the part of its orbit closest to the Sun, Pluto has an atmosphere of nitrogen and methane similar to Triton's, but these gases are frozen when it is farther from the Sun.

Other bodies within the Solar System have extremely thin atmospheres not in equilibrium. These include the Moon (sodium gas), Mercury (sodium gas), Europa (oxygen), Io (sulfur), and Enceladus (water vapor).

The first exoplanet whose atmospheric composition was determined is HD 209458b, a gas giant with a close orbit around a star in the constellation Pegasus. Its atmosphere is heated to temperatures over 1,000 K, and is steadily escaping into space. Hydrogen, oxygen, carbon and sulfur have been detected in the planet's inflated atmosphere.

ATMOSPHERE OF EARTH

The atmosphere of Earth is the layer of gases, commonly known as air that surrounds the planet Earth and is retained by Earth's gravity. The atmosphere of Earth protects life on Earth by absorbing ultraviolet solar radiation, warming the surface through heat retention (greenhouse effect), and reducing temperature extremes between day and night (the diurnal temperature variation).

By volume, dry air contains 78.09% nitrogen, 20.95% oxygen, 0.93% argon, 0.04% carbon dioxide, and small amounts of other gases. Air also contains a variable amount of water vapor, on average around 1% at sea level, and 0.4% over the entire atmosphere. Air content and atmospheric pressure vary at different layers, and air suitable for use in photosynthesis by terrestrial plants and breathing of terrestrial animals is found only in Earth's troposphere and in artificial atmospheres.

The atmosphere has a mass of about 5.15×1018 kg, three quarters of which is within about 11 km (6.8 mi; 36,000 ft) of the surface. The atmosphere becomes thinner and thinner with increasing altitude, with no definite boundary between the atmosphere and outer space. The Kármán line, at 100 km (62 mi), or 1.57% of Earth's radius, is often used as the border between the atmosphere and outer space. Atmospheric effects become noticeable during atmospheric reentry of spacecraft at an altitude of around 120 km (75 mi). Several layers can be distinguished in the atmosphere, based on characteristics such as temperature and composition.

The study of Earth's atmosphere and its processes is called atmospheric science (aerology). Early pioneers in the field include Léon Teisserenc de Bort and Richard Assmann.

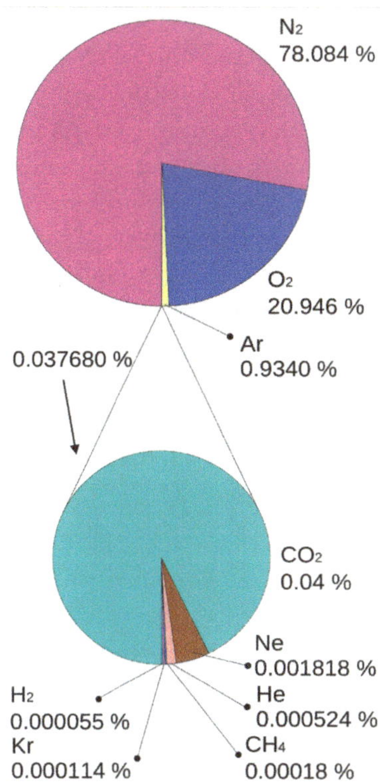

Composition of Earth's atmosphere by volume. The lower pie represents the trace gases that together compose about 0.038% of the atmosphere (0.043% with CO2 at 2014 concentration). The numbers are from a variety of years (mainly 1987, with CO2 and methane from 2009) and do not represent any single source.

Composition

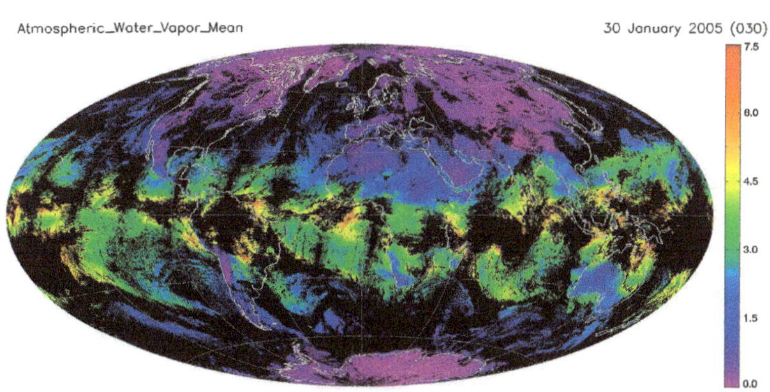

Mean atmospheric water vapor

The three major constituents of air, and therefore of Earth's atmosphere, are nitrogen, oxygen, and argon. Water vapor accounts for roughly 0.25% of the atmosphere by mass. The concentration of water vapor (a greenhouse gas) varies significantly from around 10 ppm by volume in the coldest portions of the atmosphere to as much as 5% by volume in hot, humid air masses, and concentrations of other atmospheric gases are typically quoted in terms of dry air (without water vapor). The remaining gases are often referred to as trace gases, among which are the greenhouse gases, principally carbon dioxide, methane, nitrous oxide, and ozone. Filtered air includes trace amounts of many other chemical compounds. Many substances of natural origin may be present in locally and seasonally variable small amounts as aerosols in an unfiltered air sample, including dust of mineral and organic composition, pollen and spores, sea spray, and volcanic ash. Various industrial pollutants also may be present as gases or aerosols, such as chlorine (elemental or in compounds), fluorine compounds and elemental mercury vapor. Sulfur compounds such as hydrogen sulfide and sulfur dioxide (SO_2) may be derived from natural sources or from industrial air pollution.

Major constituents of dry air, by volume			
Gas		**Volume**[(A)]	
Name	**Formula**	**in ppmv**[(B)]	**in %**
Nitrogen	N_2	780,840	78.084
Oxygen	O_2	209,460	20.946
Argon	Ar	9,340	0.9340
Carbon dioxide	CO_2	400	0.04
Neon	Ne	18.18	0.001818
Helium	He	5.24	0.000524
Methane	CH_4	1.79	0.000179
Not included in above dry atmosphere:			
Water vapor[(C)]	H_2O	10–50,000[(D)]	0.001%–5%[(D)]
notes: (A) volume fraction is equal to mole fraction for ideal gas only, also see volume (thermodynamics); (B) ppmv: parts per million by volume; (C) Water vapor is about 0.25% *by mass* over full atmosphere; (D) Water vapor strongly varies locally			

Structure of the Atmosphere

Principal Layers

In general, air pressure and density decrease with altitude in the atmosphere. However, temperature has a more complicated profile with altitude, and may remain relatively constant or even increase with altitude in some regions (see the temperature section, below). Because the general pattern of the temperature/altitude profile is constant and measurable by means of instrumented balloon soundings, the temperature behavior provides a useful metric to distinguish atmospheric layers. In this way, Earth's atmosphere can be divided (called atmospheric stratification) into five main layers. Excluding the exosphere, Earth has four primary layers, which are the troposphere, stratosphere, mesosphere, and thermosphere. From highest to lowest, the five main layers are:

- Exosphere: 700 to 10,000 km (440 to 6,200 miles)
- Thermosphere: 80 to 700 km (50 to 440 miles)
- Mesosphere: 50 to 80 km (31 to 50 miles)
- Stratosphere: 12 to 50 km (7 to 31 miles)
- Troposphere: 0 to 12 km (0 to 7 miles)

Exosphere

The exosphere is the outermost layer of Earth's atmosphere (i.e. the upper limit of the atmosphere). It extends from the exobase, which is located at the top of the thermosphere at an altitude of about 700 km above sea level, to about 10,000 km (6,200 mi; 33,000,000 ft) where it merges into the solar wind.

This layer is mainly composed of extremely low densities of hydrogen, helium and several heavier molecules including nitrogen, oxygen and carbon dioxide closer to the exobase. The atoms and molecules are so far apart that they can travel hundreds of kilometers without colliding with one another. Thus, the exosphere no longer behaves like a gas, and the particles constantly escape into space. These free-moving particles follow ballistic trajectories and may migrate in and out of the magnetosphere or the solar wind.

The exosphere is located too far above Earth for any meteorological phenomena to be possible. However, the aurora borealis and aurora australis sometimes occur in the lower part of the exosphere, where they overlap into the thermosphere. The exosphere contains most of the satellites orbiting Earth.

Earth's Exosphere

The most common molecules within Earth's exosphere are those of the lightest atmospheric gasses. Hydrogen is present throughout the exosphere, with some helium, carbon dioxide, and atomic oxygen near its base. Because it can be difficult to define the boundary between the exosphere and outer space (see "Upper boundary" at the end of this section), the exosphere may be considered a part of interplanetary or outer space.

Lower Boundary

The lower boundary of the exosphere is called the *exobase*. It is also called *exopause* and *‹critical altitude'* as this is the altitude where barometric conditions no longer apply. Atmospheric temperature becomes nearly a constant above this altitude. On Earth, the altitude of the exobase ranges from about 500 to 1,000 kilometres (310 to 620 mi) depending on solar activity.

The exobase can be defined in one of two ways:

If we define the exobase as the height at which upward-traveling molecules experience one collision on average, then at this position the mean free path of a molecule is equal to one pressure scale height. This is shown in the following. Consider a volume of air, with horizontal area A and height equal to the mean free path I at pressure p and temperature T For an ideal gas, the number of molecules contained in it is:

$$n = \frac{pAl}{RT}$$

where R is the universal gas constant. From the requirement that each molecule traveling upward undergoes on average one collision, the pressure is:

$$p = \frac{m_A n g}{A}$$

where m_A is the mean molecular mass of the gas. Solving these two equations gives:

$$l = \frac{RT}{m_A g}$$

which is the equation for the pressure scale height. As the pressure scale height is almost equal to the density scale height of the primary constituent, and because the Knudsen number is the ratio of mean free path and typical density fluctuation scale, this means that the exobase lies in the region where $\mathrm{Kn}(h_{EB}) \simeq 1$

The fluctuation in the height of the exobase is important because this provides atmospheric drag on satellites, eventually causing them to fall from orbit if no action is taken to maintain the orbit.

Upper Boundary of Earth

In principle, the exosphere covers distances where particles are still gravitationally bound to Earth, i.e. particles still have ballistic orbits that will take them back towards Earth. The upper boundary of the exosphere can be defined as the distance at which the influence of solar radiation pressure on atomic hydrogen exceeds that of Earth's gravitational pull. This happens at half the distance to the Moon (the average distance between Earth and the Moon is 384,400 kilometres (238,900 mi)). The exosphere, observable from space as the geocorona, is seen to extend to at least 10,000 kilometres (6,200 mi) from Earth's surface. The exosphere is a transitional zone between Earth's atmosphere and space.

Moon's Exosphere

On 17 August 2015, based on studies with the Lunar Atmosphere and Dust Environment Explorer (LADEE) spacecraft, NASA scientists reported the detection of neon in the exosphere of the moon.

Thermosphere

The thermosphere is the layer of the Earth's atmosphere directly above the mesosphere. The exosphere is above that but is a minor layer of the atmosphere. Within this layer of the atmosphere, ultraviolet radiation causes photoionization/photodissociation of molecules, creating ions in the ionosphere. Taking its name from the Greek θερμός (pronounced *thermos*) meaning heat, the thermosphere begins about 85 kilometres (53 mi) above the Earth. At these high altitudes, the residual atmospheric gases sort into strata according to molecular mass (see turbosphere). Thermospheric temperatures increase with altitude due to absorption of highly energetic solar radiation. Temperatures are highly dependent on solar activity, and can rise to 2,000 °C (3,630 °F). Radiation causes the atmosphere particles in this layer to become electrically charged (see ionosphere), enabling radio waves to be refracted and thus be received beyond the horizon. In the exosphere, beginning at 500 to 1,000 kilometres (310 to 620 mi) above the Earth's surface, the atmosphere turns into space.

The highly diluted gas in this layer can reach 2,500 °C (4,530 °F) during the day. Even though the temperature is so high, one would not feel warm in the thermosphere, because it is so near vacuum that there is not enough contact with the few atoms of gas to transfer much heat. A normal thermometer might be significantly below 0 °C (32 °F), at least at night, because the energy lost by thermal radiation would exceed the energy acquired from the atmospheric gas by direct contact. In the anacoustic zone above 160 kilometres (99 mi), the density is so low that molecular interactions are too infrequent to permit the transmission of sound.

The dynamics of the thermosphere are dominated by atmospheric tides, which are driven by the very significant diurnal heating. Atmospheric waves dissipate above this level because of collisions between the neutral gas and the ionospheric plasma.

The International Space Station orbits the Earth within the middle of the thermosphere, between 330 and 435 kilometres (205 and 270 mi) (decaying by 2 km/month and raised by periodic reboosts), whereas the Gravity Field and Steady-State Ocean Circulation Explorer satellite at 260 kilometres (160 mi) utilized winglets and an innovative ion engine to maintain a stable orientation and orbit.

Neutral Gas Constituents

It is convenient to separate the atmospheric regions according to the two temperature minima at about 12 km altitude (the tropopause) and at about 85 km (the mesopause). The thermosphere (or the upper atmosphere) is the height region above 85 km, while the region between the tropopause and the mesopause is the middle atmosphere (stratosphere and mesosphere) where absorption of solar UV radiation generates the temperature maximum near 45 km altitude and causes the ozone layer.

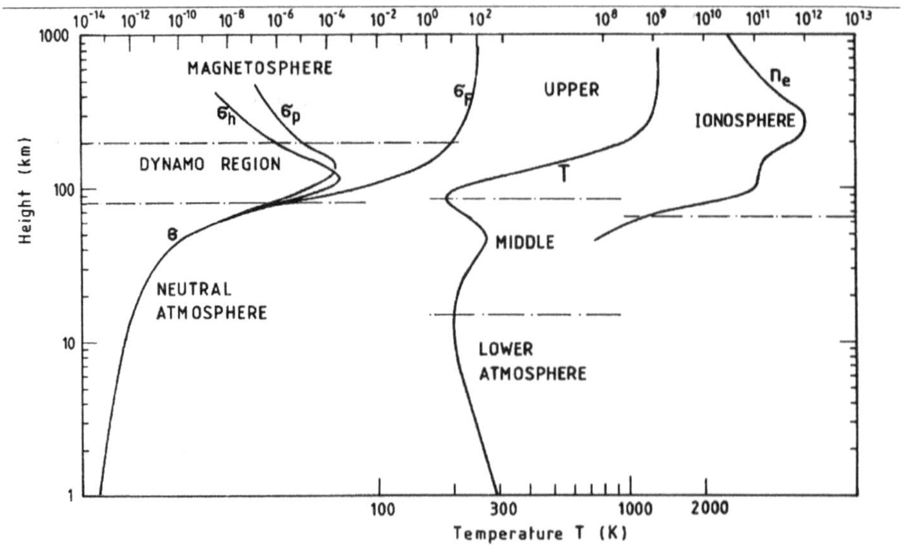

Nomenclature of atmospheric regions based on the profiles of electric conductivity (left), temperature (middle), and electron density (right)

The density of the Earth's atmosphere decreases nearly exponentially with altitude. The total mass of the atmosphere is $M = \rho_A H \simeq 1$ kg/cm^2 within a column of one square centimeter above the ground (with $\rho_A = 1.29$ kg/m^3 the atmospheric density on the ground at $z = 0$ m altitude, and $H \simeq 8$ km the average atmospheric scale height). 80% of that mass is concentrated within the troposphere. The mass of the thermosphere above about 85 km is only 0.002% of the total mass. Therefore, no significant energetic feedback from the thermosphere to the lower atmospheric regions can be expected.

Turbulence causes the air within the lower atmospheric regions below the turbopause at about 110 km to be a mixture of gases that does not change its composition. Its mean molecular weight is 29 g/mol with molecular oxygen (O_2) and nitrogen (N_2) as the two dominant constituents. Above the turbopause, however, diffusive separation of the various constituents is significant, so that each constituent follows its own barometric height structure with a scale height inversely proportional to its molecular weight. The lighter constituents atomic oxygen (O), helium (He), and hydrogen (H) successively dominate above about 200 km altitude and vary with geographic location, time, and solar activity. The ratio N_2/O which is a measure of the electron density at the ionospheric F region is highly affected by these variations. These changes follow from the diffusion of the minor constituents through the major gas component during dynamic processes.

The thermosphere contains an appreciable concentration of elemental sodium located in a 10-km thick band that occurs at the edge of the mesosphere, 80 to 100 km above Earth's surface. The sodium has an average concentration of 400,000 atoms per cubic centimeter. This band is regularly replenished by sodium sublimating from incoming meteors. Astronomers have begun utilizing this sodium band to create "guide stars" as part of the optical correction process in producing ultra-sharp ground-based observations.

Energy Input

Energy Budget

The thermospheric temperature can be determined from density observations as well as from direct satellite measurements. The temperature vs. altitude z can be simulated by the so-called Bates profile:

$$T = T_\infty - (T_\infty - T_0)\exp\{-s(z - z_0)\}$$

with T_∞ the exospheric temperature above about 400 km altitude, T_o = 355 K, and z_o = 120 km reference temperature and height, and s an empirical parameter depending on T_∞ and decreasing with T_∞. That formula is derived from a simple equation of heat conduction. One estimates a total heat input of $q_o \simeq$ 0.8 to 1.6 mW/m^2 above z_o = 120 km altitude. In order to obtain equilibrium conditions, that heat input q_o above z_o is lost to the lower atmospheric regions by heat conduction.

The exospheric temperature T_∞ is a fair measurement of the solar XUV radiation. Since solar radio emission F at 10.7 cm wavelength is a good indicator of solar activity, one can apply the empirical formula for quiet magnetospheric conditions.

$$T_\infty \simeq 500 + 3.4F_0$$

with T_∞ in K, F_o in 10^{-2} W m^{-2} Hz^{-1} (the Covington index) a value of F averaged over several solar cycles. The Covington index varies typically between 70 and 250 during a solar cycle, and never drops below about 50. Thus, T_∞ varies between about 740 and 1350 K. During very quiet magnetospheric conditions, the still continuously flowing magnetospheric energy input contributes by about 250 K to the residual temperature of 500 K in eq.(2). The rest of 250 K in eq.(2) can be attributed to atmospheric waves generated within the troposphere and dissipated within the lower thermosphere.

Solar XUV Radiation

The solar X-ray and extreme ultraviolet radiation (XUV) at wavelengths < 170 nm is almost completely absorbed within the thermosphere. This radiation causes the various ionospheric layers as well as a temperature increase at these heights. While the solar visible light (380 to 780 nm) is nearly constant with a variability of not more than about 0.1% of the solar constant, the solar XUV radiation is highly variable in time and space. For instance, X-ray bursts associated with solar flares can dramatically increase their intensity over preflare levels by many orders of magnitude over a time span of tens of minutes. In the extreme ultraviolet, the Lyman α line at 121.6 nm represents an important source of ionization and dissociation at ionospheric D layer heights. During quiet periods of solar activity, it alone contains more energy than the rest of the XUV spectrum. Quasi-periodic changes of the order of 100% or greater, with periods with period of 27 days and 11 years, belong to the prominent variations of solar XUV radiation. However, irregular fluctuations over all time scales are present all the time. During low solar activity, about half of the total energy input into the thermosphere is thought to be solar XUV radiation. Evidently, that solar XUV energy input occurs only during daytime conditions, maximizing at the equator during equinox.

Solar Wind

A second source of energy input into the thermosphere is solar wind energy which is transferred to the magnetosphere by mechanisms that are not well understood. One possible way to transfer energy is via a hydrodynamic dynamo process. Solar wind particles penetrate into the polar regions of the magnetosphere where the geomagnetic field lines are essentially vertically directed. An electric field is generated, directed from dawn to dusk. Along the last closed geomagnetic field lines with their footpoints within the auroral zones, field aligned electric currents can flow into the ionospheric dynamo region where they are closed by electric Pedersen and Hall currents. Ohmic losses of the Pedersen currents heat the lower thermosphere (see e.g., Magnetospheric electric convection field). In addition, penetration of high energetic particles from the magnetosphere into the auroral regions enhance drastically the electric conductivity, further increasing the electric currents and thus Joule heating. During quiet magnetospheric activity, the magnetosphere contributes perhaps by a quarter to the thermosphere's energy budget. This is about 250 K of the exospheric temperature in eq.(2). During very large activity, however, this heat input can increase substantially, by a factor of four or more. That solar wind input occurs mainly in the auroral regions during both day and night.

Atmospheric Waves

Two kinds of large-scale atmospheric waves within the lower atmosphere exist: internal waves with finite vertical wavelengths which can transport wave energy upward; and external waves with infinitely large wavelengths which cannot transport wave energy. Atmospheric gravity waves and most of the atmospheric tides generated within the troposphere belong to the internal waves. Their density amplitudes increase exponentially with height, so that at the mesopause these waves become turbulent and their energy is dissipated (similar to breaking of ocean waves at the coast), thus contributing to the heating of the thermosphere by about 250 K in eq.(2). On the other hand, the fundamental diurnal tide labelled (1, −2) which is most efficiently excited by solar irradiance is an external wave and plays only a marginal role within lower and middle atmosphere. However, at thermospheric altitudes, it becomes the predominant wave. It drives the electric Sq-current within the ionospheric dynamo region between about 100 and 200 km height.

Heating, predominately by tidal waves, occurs mainly at lower and middle latitudes. The variability of this heating depends on the meteorological conditions within troposphere and middle atmosphere, and may not exceed about 50%.

Dynamics

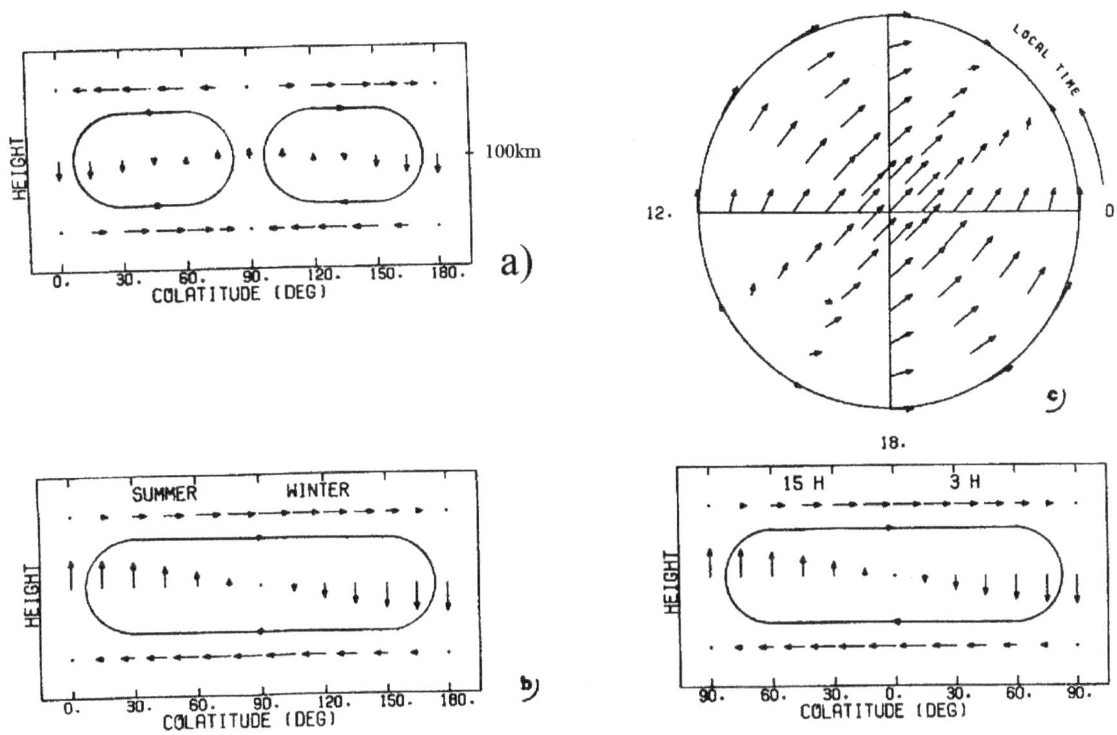

Schematic meridian-height cross-section of circulation of (a) symmetric wind component (P_2^0), (b) of antisymmetric wind component (P_1^0), and (d) of symmetric diurnal wind component (P_1^1) at 3 h and 15 h local time. Upper right pannel (c) shows the horizontal wind vectors of the diurnal component in the northern hemisphere depending on local time.

Within the thermosphere above about 150 km height, all atmospheric waves successively become external waves, and no significant vertical wave structure is visible. The atmospheric wave modes degenerate to the spherical functions P_n^m with m a meridional wave number and n the zonal wave number (m = 0: zonal mean flow; m = 1: diurnal tides; m = 2: semidiurnal tides; etc.). The thermosphere becomes a damped oscillator system with low-pass filter characteristics. This means that smaller-scale waves (greater numbers of (n,m)) and higher frequencies are suppressed in favor of large-scale waves and lower frequencies.

If one considers very quiet magnetospheric disturbances and a constant mean exospheric temperature (averaged over the sphere), the observed temporal and spatial distribution of the exospheric temperature distribution can be described by a sum of spheric functions:

$$T(\varphi,\lambda,t) = T_\infty\{1 + \Delta T_2^0 P_2^0(\varphi) + \Delta T_1^0 P_1^0(\varphi)\cos[\omega_a(t-t_a)] + \Delta T_1^1 P_1^1(\varphi)\cos(\tau-\tau_d) + \ldots\}$$

Here, it is φ latitude, λ longitude, and t time, ω_a the angular frequency of one year, ω_d the angular frequency of one solar day, and $\tau = \omega_d t + \lambda$ the local time. t_a = June 21 is the date of northern summer solstice, and τ_d = 15:00 is the local time of maximum diurnal temperature.

The first term in (3) on the right is the global mean of the exospheric temperature (of the order of 1000 K). The second term with $P_2^0 = 0.5(3\sin^2(\varphi)-1)$ represents the heat surplus at lower latitudes and a corresponding heat deficit at higher latitudes. A thermal wind system develops with wind toward the poles in the upper level and wind away from the poles in the lower level. The coefficient $\Delta T_2^0 \approx 0.004$ is small because Joule heating in the aurora regions compensates that heat surplus even during quiet magnetospheric conditions. During disturbed conditions, however, that term becomes dominant, changing sign so that now heat surplus is transported from the poles to the equator. The third term (with $P_1^0 = \sin\varphi$) represents heat surplus on the summer hemisphere and is responsible for the transport of excess heat from the summer into the winter hemisphere. Its relative amplitude is of the order $\Delta T_1^0 \simeq 0.13$. The fourth term (with $P_1^1(\varphi) = \cos\varphi$) is the dominant diurnal wave (the tidal mode (1,−2)). It is responsible for the transport of excess heat from the daytime hemisphere into the nighttime hemisphere. Its relative amplitude is $\Delta T_1^1 \simeq 0.15$, thus on the order of 150 K. Additional terms (e.g., semiannual, semidiurnal terms and higher order terms) must be added to eq.(3). However, they are of minor importance. Corresponding sums can be developed for density, pressure, and the various gas constituents.

Thermospheric Storms

In contrast to solar XUV radiation, magnetospheric disturbances, indicated on the ground by geomagnetic variations, show an unpredictable impulsive character, from short periodic disturbances of the order of hours to long-standing giant storms of several day's duration. The reaction of the thermosphere to a large magnetospheric storm is called thermospheric storm. Since the heat input into the thermosphere occurs at high latitudes (mainly into the auroral regions), the heat transport represented by the term P_2^0 in eq.(3) is reversed. In addition, due to the impulsive form of the disturbance, higher-order terms are generated which, however, possess short decay times and thus quickly disappear. The sum of these modes determines the travel time of the disturbance to the lower latitudes, and thus the response time of the thermosphere with respect to the magnetospheric disturbance. Important for the development of an ionospheric storm is the increase of the ratio N_2/O during a thermospheric storm at middle and higher latitude. An increase of N_2 increases the loss process of the ionospheric plasma and causes therefore a decrease of the electron density within the ionospheric F-layer (negative ionospheric storm).

Mesosphere

The **mesosphere** is the layer of the Earth's atmosphere that is directly above the stratosphere and directly below the mesopause. In the mesosphere, temperature decreases as the altitude increases. The upper boundary of the mesosphere is the mesopause, which can be the coldest naturally occurring place on Earth with temperatures below −143 °C (−225 °F; 130 K). The exact upper and lower boundaries of the mesosphere vary with latitude and with season, but the lower boundary of the mesosphere is usually located at heights of about 50 kilometres (160,000 ft; 31 mi) above the Earth's surface and the mesopause is usually at heights near 100 kilometres (62 mi), except at middle and high latitudes in summer where it descends to heights of about 85 kilometres (53 mi; 279,000 ft).

The stratosphere, mesosphere and lowest part of the thermosphere are collectively referred to as the middle atmosphere, which spans heights from approximately 10 kilometres (33,000 ft; 6.2 mi) to 100 kilometres (62 mi; 330,000 ft). The mesopause, at an altitude of 80–90 km (50–56 mi), separates the mesosphere from the thermosphere—the second-outermost layer of the Earth's atmosphere. This is also around the same altitude as the turbopause, below which different chemical species are well mixed due to turbulent eddies. Above this level the atmosphere becomes non-uniform; the scale heights of different chemical species differ by their molecular masses.

Space Shuttle Endeavour appears to straddle the stratosphere and mesosphere in this photo. The troposphere, which contains clouds, appears orange in this photo.

Temperature

Within the mesosphere, temperature decreases with increasing height, due to decreasing solar heating and increasing cooling by CO_2 radiative emission. The top of the mesosphere, called the mesopause, is the coldest part of Earth's atmosphere. Temperatures in the upper mesosphere fall as low as −101 °C (172 K; −150 °F),varying according to latitude and season.

Dynamic Features

The main dynamic features in this region are strong zonal (East-West) winds, atmospheric tides, internal atmospheric gravity waves (commonly called "gravity waves"), and planetary waves. Most of these tides and waves start in the troposphere and lower stratosphere, and propagate to the mesosphere. In the mesosphere, gravity-wave amplitudes can become so large that the waves become unstable and dissipate. This dissipation deposits momentum into the mesosphere and largely drives global circulation.

Noctilucent clouds are located in the mesosphere. The upper mesosphere is also the region of the ionosphere known as the *D layer*. The D layer is only present during the day, when some ionization occurs with nitric oxide being ionized by Lyman series-alpha hydrogen radiation. The ionization is so weak that when night falls, and the source of ionization is removed, the free electron and ion form back into a neutral molecule. The mesosphere has been called the "ignorosphere" because it is poorly studied relative to the stratosphere (which can be accessed with high-altitude balloons) and the thermosphere (in which satellites can orbit). A 5 km (3.1 mi; 16,000 ft) deep sodium layer is located between 80–105 km (50–65 mi; 262,000–344,000 ft). Made of unbound, non-ionized atoms of sodium, the sodium layer radiates weakly to contribute to the airglow. The sodium has an average concentration of 400,000 atoms per cubic centimeter. This band is regularly replenished by sodium sublimating from incoming meteors. Astronomers have begun utilizing this sodium band to create "guide stars" as part of the adaptive optical correction process used to produce ultra-sharp ground-based observations.

Millions of meteors enter the Earth's atmosphere, averaging 40 tons per year.

Uncertainties

The mesosphere lies above altitude records for aircraft and balloons, and below the minimum altitude for orbital spacecraft. It has only been accessed through the use of sounding rockets. As a result, it is the least-understood part of the atmosphere. The presence of red sprites and blue jets (electrical discharges or lightning within the lower mesosphere), noctilucent clouds, and density shears within this poorly understood layer are of current scientific interest.

Stratosphere

The stratosphere is the second-lowest layer of Earth's atmosphere. It lies above the troposphere and is separated from it by the tropopause. This layer extends from the top of the troposphere at roughly 12 km (7.5 mi; 39,000 ft) above Earth's surface to the stratopause at an altitude of about 50 to 55 km (31 to 34 mi; 164,000 to 180,000 ft).

The atmospheric pressure at the top of the stratosphere is roughly 1/1000 the pressure at sea level. It contains the ozone layer, which is the part of Earth's atmosphere that contains relatively high concentrations of that gas. The stratosphere defines a layer in which temperatures rise with increasing altitude. This rise in temperature is caused by the absorption of ultraviolet radiation (UV) radiation from the Sun by the ozone layer, which restricts turbulence and mixing. Although the temperature may be −60 °C (−76 °F; 210 K) at the tropopause, the top of the stratosphere is much warmer, and may be near 0 °C.

The stratospheric temperature profile creates very stable atmospheric conditions, so the stratosphere lacks the weather-producing air turbulence that is so prevalent in the troposphere. Consequently, the stratosphere is almost completely free of clouds and other forms of weather. However, polar stratospheric or nacreous clouds are occasionally seen in the lower part of this layer of the atmosphere where the air is coldest. The stratosphere is the highest layer that can be accessed by jet-powered aircraft.

Troposphere

The troposphere is the lowest layer of Earth's atmosphere. It extends from Earth's surface to an average height of about 12 km, although this altitude actually varies from about 9 km (30,000 ft) at the poles to 17 km (56,000 ft) at the equator, with some variation due to weather. The troposphere is bounded above by the tropopause, a boundary marked in most places by a temperature inversion (i.e. a layer of relatively warm air above a colder one), and in others by a zone which is isothermal with height.

Although variations do occur, the temperature usually declines with increasing altitude in the troposphere because the troposphere is mostly heated through energy transfer from the surface. Thus, the lowest part of the troposphere (i.e. Earth's surface) is typically the warmest section of the troposphere. This promotes vertical mixing (hence the origin of its name in the Greek word τρόπος, *tropos*, meaning "turn"). The troposphere contains roughly 80% of the mass of Earth's atmosphere. The troposphere is denser than all its overlying atmospheric layers because a larger atmospheric weight sits on top of the troposphere and causes it to be most severely compressed. Fifty percent of the total mass of the atmosphere is located in the lower 5.6 km (18,000 ft) of the troposphere.

Nearly all atmospheric water vapor or moisture is found in the troposphere, so it is the layer where most of Earth's weather takes place. It has basically all the weather-associated cloud genus types generated by active wind circulation, although very tall cumulonimbus thunder clouds can penetrate the tropopause from below and rise into the lower part of the stratosphere. Most conventional aviation activity takes place in the troposphere, and it is the only layer that can be accessed by propeller-driven aircraft.

Other Layers

Within the five principal layers that are largely determined by temperature, several secondary layers may be distinguished by other properties:

- The ozone layer is contained within the stratosphere. In this layer ozone concentrations are about 2 to 8 parts per million, which is much higher than in the lower atmosphere but still very small compared to the main components of the atmosphere. It is mainly located in the lower portion of the stratosphere from about 15–35 km (9.3–21.7 mi; 49,000–115,000 ft), though the thickness varies seasonally and geographically. About 90% of the ozone in Earth's atmosphere is contained in the stratosphere.

- The ionosphere is a region of the atmosphere that is ionized by solar radiation. It is responsible for auroras. During daytime hours, it stretches from 50 to 1,000 km (31 to 621 mi; 160,000 to 3,280,000 ft) and includes the mesosphere, thermosphere, and parts of the exosphere. However, ionization in the mesosphere largely ceases during the night, so auroras are normally seen only in the thermosphere and lower exosphere. The ionosphere forms the inner edge of the magnetosphere. It has practical importance because it influences, for example, radio propagation on Earth.
- The homosphere and heterosphere are defined by whether the atmospheric gases are well mixed. The surface-based homosphere includes the troposphere, stratosphere, mesosphere, and the lowest part of the thermosphere, where the chemical composition of the atmosphere does not depend on molecular weight because the gases are mixed by turbulence. This relatively homogeneous layer ends at the *turbopause* found at about 100 km (62 mi; 330,000 ft), the very edge of space itself as accepted by the FAI, which places it about 20 km (12 mi; 66,000 ft) above the mesopause.

Above this altitude lies the heterosphere, which includes the exosphere and most of the thermosphere. Here, the chemical composition varies with altitude. This is because the distance that particles can move without colliding with one another is large compared with the size of motions that cause mixing. This allows the gases to stratify by molecular weight, with the heavier ones, such as oxygen and nitrogen, present only near the bottom of the heterosphere. The upper part of the heterosphere is composed almost completely of hydrogen, the lightest element.

- The planetary boundary layer is the part of the troposphere that is closest to Earth's surface and is directly affected by it, mainly through turbulent diffusion. During the day the planetary boundary layer usually is well-mixed, whereas at night it becomes stably stratified with weak or intermittent mixing. The depth of the planetary boundary layer ranges from as little as about 100 metres (330 ft) on clear, calm nights to 3,000 m (9,800 ft) or more during the afternoon in dry regions.

The average temperature of the atmosphere at Earth's surface is 14 °C (57 °F; 287 K) or 15 °C (59 °F; 288 K), depending on the reference.

Physical Properties

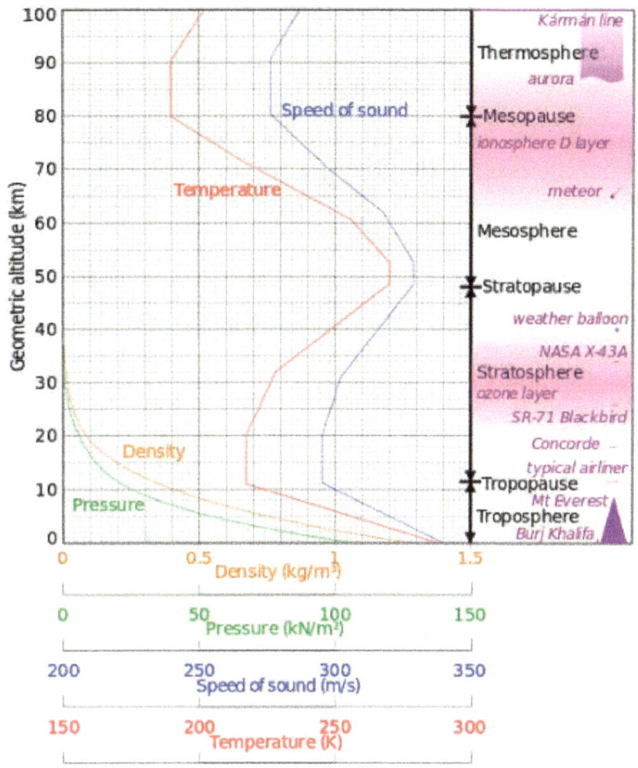

Comparison of the 1962 US Standard Atmosphere graph of geometric altitude against air density, pressure, the speed of sound and temperature with approximate altitudes of various objects.

Pressure and Thickness

The average atmospheric pressure at sea level is defined by the International Standard Atmosphere as 101325 pascals (760.00 Torr; 14.6959 psi; 760.00 mmHg). This is sometimes referred to as a unit of standard atmospheres (atm). Total atmospheric mass is 5.1480×10^{18} kg (1.135×10^{19} lb), about 2.5% less than would be inferred from the average sea level pressure and Earth's area of 51007.2 megahectares, this portion being displaced by Earth's mountainous terrain. Atmospheric pressure is the total weight of the air above unit area at the point where the pressure is measured. Thus air pressure varies with location and weather.

If the entire mass of the atmosphere had a uniform density from sea level, it would terminate abruptly at an altitude of 8.50 km (27,900 ft). It actually decreases exponentially with altitude, dropping by half every 5.6 km (18,000 ft) or by a factor of 1/e every 7.64 km (25,100 ft), the average scale height of the atmosphere below 70 km (43 mi; 230,000 ft). However, the atmosphere is more accurately modeled with a customized equation for each layer that takes gradients of temperature, molecular composition, solar radiation and gravity into account.

In summary, the mass of Earth's atmosphere is distributed approximately as follows:

- 50% is below 5.6 km (18,000 ft).
- 90% is below 16 km (52,000 ft).
- 99.99997% is below 100 km (62 mi; 330,000 ft), the Kármán line. By international convention, this marks the beginning of space where human travelers are considered astronauts.

By comparison, the summit of Mt. Everest is at 8,848 m (29,029 ft); commercial airliners typically cruise between 10 km (33,000 ft) and 13 km (43,000 ft) where the thinner air improves fuel economy; weather balloons reach 30.4 km (100,000 ft) and above; and the highest X-15 flight in 1963 reached 108.0 km (354,300 ft).

Even above the Kármán line, significant atmospheric effects such as auroras still occur. Meteors begin to glow in this region, though the larger ones may not burn up until they penetrate more deeply. The various layers of Earth's ionosphere, important to HF radio propagation, begin below 100 km and extend beyond 500 km. By comparison, the International Space Station and Space Shuttle typically orbit at 350–400 km, within the F-layer of the ionosphere where they encounter enough atmospheric drag to require reboosts every few months. Depending on solar activity, satellites can experience noticeable atmospheric drag at altitudes as high as 700–800 km.

Temperature and Speed of Sound

The division of the atmosphere into layers mostly by reference to temperature is discussed above. Temperature decreases with altitude starting at sea level, but variations in this trend begin above 11 km, where the temperature stabilizes through a large vertical distance through the rest of the troposphere. In the stratosphere, starting above about 20 km, the temperature increases with height, due to heating within the ozone layer caused by capture of significant ultraviolet radiation from the Sun by the dioxygen and ozone gas in this region. Still another region of increasing temperature with altitude occurs at very high altitudes, in the aptly-named thermosphere above 90 km.

Because in an ideal gas of constant composition the speed of sound depends only on temperature and not on the gas pressure or density, the speed of sound in the atmosphere with altitude takes on the form of the complicated temperature profile (illustration to the right), and does not mirror altitudinal changes in density or pressure.

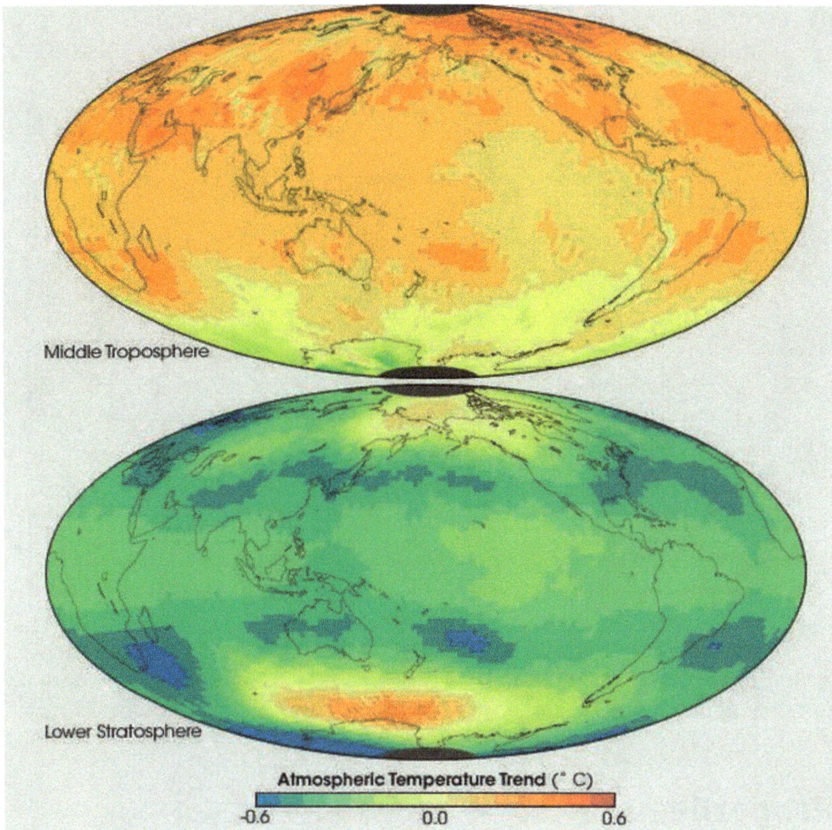

These images show temperature trends in two thick layers of the atmosphere as measured by a series of satellite-based instruments between January 1979 and December 2005. The measurements were taken by Microwave Sounding Units and Advanced Microwave Sounding Units flying on a series of National Oceanic and Atmospheric Administration (NOAA) weather satellites. The instruments record microwaves emitted from oxygen molecules in the atmosphere. Source:

Density and Mass

The density of air at sea level is about 1.2 kg/m^3 (1.2 g/L, 0.0012 g/cm^3). Density is not measured directly but is calculated from measurements of temperature, pressure and humidity using the equation of state for air (a form of the ideal gas law). Atmospheric density decreases as the altitude increases. This variation can be approximately modeled using the barometric formula. More sophisticated models are used to predict orbital decay of satellites.

The average mass of the atmosphere is about 5 quadrillion (5×10^{15}) tonnes or 1/1,200,000 the mass of Earth. According to the American National Center for Atmospheric Research, "The total mean mass of the atmosphere is 5.1480×10^{18} kg with an annual range due to water vapor of 1.2 or 1.5×10^{15} kg, depending on whether surface pressure or water vapor data are used; somewhat smaller than the previous estimate. The mean mass of water vapor is estimated as 1.27×10^{16} kg and the dry air mass as $5.1352 \pm 0.0003\times10^{18}$ kg."

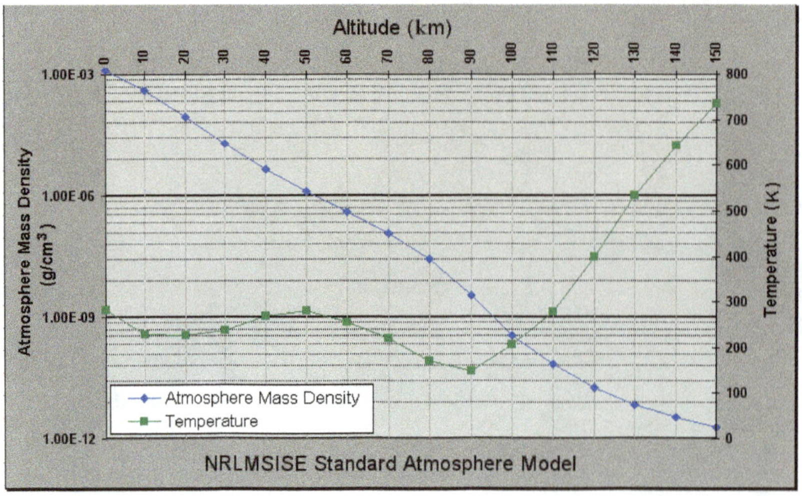

Temperature and mass density against altitude from the NRLMSISE-00standard atmosphere model (the eight dotted lines in each "decade" are at the eight cubes 8, 27, 64, ..., 729)

Optical Properties

Solar radiation (or sunlight) is the energy Earth receives from the Sun. Earth also emits radiation back into space, but at longer wavelengths that we cannot see. Part of the incoming and emitted radiation is absorbed or reflected by the atmosphere.

Scattering

When light passes through Earth's atmosphere, photons interact with it through *scattering*. If the light does not interact with the atmosphere, it is called *direct radiation* and is what you see if you were to look directly at the Sun. *Indirect radiation* is light that has been scattered in the atmosphere. For example, on an overcast day when you cannot see your shadow there is no direct radiation reaching you, it has all been scattered. As another example, due to a phenomenon called Rayleigh scattering, shorter (blue) wavelengths scatter more easily than longer (red) wavelengths. This is why the sky looks blue; you are seeing scattered blue light. This is also why sunsets are red. Because the Sun is close to the horizon, the Sun's rays pass through more atmosphere than normal to reach your eye. Much of the blue light has been scattered out, leaving the red light in a sunset.

Absorption

Different molecules absorb different wavelengths of radiation. For example, O_2 and O_3 absorb almost all wavelengths shorter than 300 nanometers. Water (H_2O) absorbs many wavelengths above 700 nm. When a molecule absorbs a photon, it increases the energy of the molecule. This heats the atmosphere, but the atmosphere also cools by emitting radiation, as discussed below.

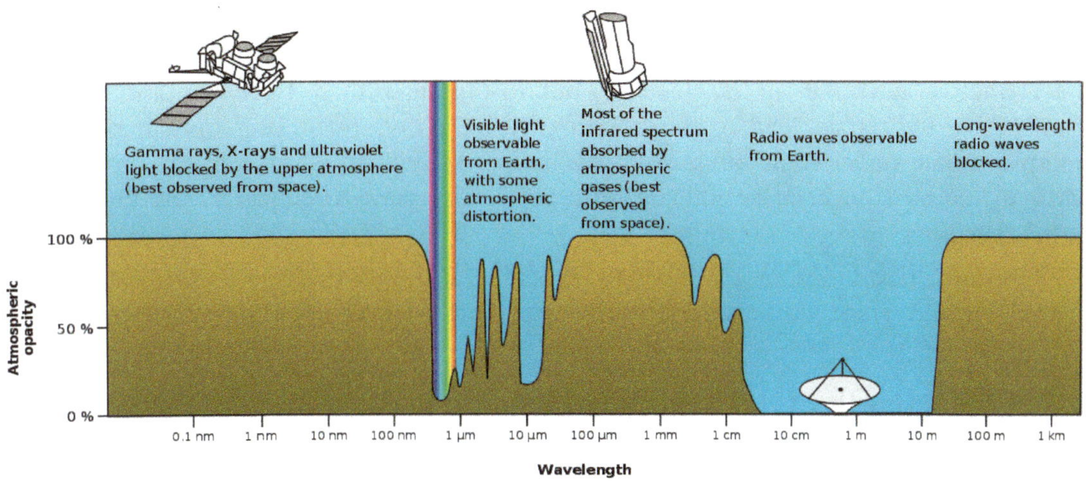

Rough plot of Earth's atmospheric transmittance (or opacity) to various wavelengths of electromagnetic radiation, including visible light.

The combined absorption spectra of the gases in the atmosphere leave "windows" of low opacity, allowing the transmission of only certain bands of light. The optical window runs from around 300 nm (ultraviolet-C) up into the range humans can see, the visible spectrum (commonly called light), at roughly 400–700 nm and continues to the infrared to around 1100 nm. There are also infrared and radio windows that transmit some infrared and radio waves at longer wavelengths. For example, the radio window runs from about one centimeter to about eleven-meter waves.

Emission

Emission is the opposite of absorption, it is when an object emits radiation. Objects tend to emit amounts and wavelengths of radiation depending on their "black body" emission curves, therefore hotter objects tend to emit more radiation, with shorter wavelengths. Colder objects emit less radiation, with longer wavelengths. For example, the Sun is approximately 6,000 K (5,730 °C; 10,340 °F), its radiation peaks near 500 nm, and is visible to the human eye. Earth is approximately 290 K (17 °C; 62 °F), so its radiation peaks near 10,000 nm, and is much too long to be visible to humans.

Because of its temperature, the atmosphere emits infrared radiation. For example, on clear nights Earth's surface cools down faster than on cloudy nights. This is because clouds (H_2O) are strong absorbers and emitters of infrared radiation. This is also why it becomes colder at night at higher elevations.

The greenhouse effect is directly related to this absorption and emission effect. Some gases in the atmosphere absorb and emit infrared radiation, but do not interact with sunlight in the visible spectrum. Common examples of these are CO_2 and H_2O.

Refractive Index

The refractive index of air is close to, but just greater than 1. Systematic variations in refractive index can lead to the bending of light rays over long optical paths. One example is that, under some circumstances, observers onboard ships can see other vessels just over the horizon because light is refracted in the same direction as the curvature of Earth's surface.

The refractive index of air depends on temperature, giving rise to refraction effects when the temperature gradient is large. An example of such effects is the mirage.

ATMOSPHERIC CIRCULATION

Atmospheric circulation is the large-scale movement of air, and together with ocean circulation is the means by which thermal energy is redistributed on the surface of the Earth. The Earth's atmospheric circulation varies from year to year, but the large scale structure of its circulation remains fairly constant. The smaller scale weather systems – mid-latitude depressions, or tropical convective cells – occur randomly, and long range weather predictions of those cannot be made beyond ten days in practice, or a month in theory (see Chaos theory and Butterfly effect). The large scale atmospheric circulation of the Earth, however, is an average of its systems and patterns, and is considered stable over longer periods of time.

The Earth's weather is a consequence of its illumination by the Sun, and the laws of thermodynamics. The atmospheric circulation can be viewed, from that standpoint, as a heat engine driven by the Sun's energy, and whose energy sink, ultimately, is the blackness of space. The work produced by that engine causes the motion of the masses of air and in that process it redistributes the energy absorbed by the Earth's surface near the tropics to space and incidentally to the latitudes nearer the poles.

The large scale atmospheric circulation cells shift polewards in warmer periods (e.g. interglacials compared to glacials), but remain largely constant as they are, fundamentally, a property of the Earth's size, rotation rate, heating and atmospheric depth, all of which change little. Over very long time periods (hundreds of millions of years), a tectonic uplift can significantly alter their major elements, such as the jet stream, and plate tectonics may shift ocean currents. During the extremely hot climates of the Mesozoic, a third desert belt may have existed at the Equator. But, the overall latitudinal pattern of Earth's climate has not changed.

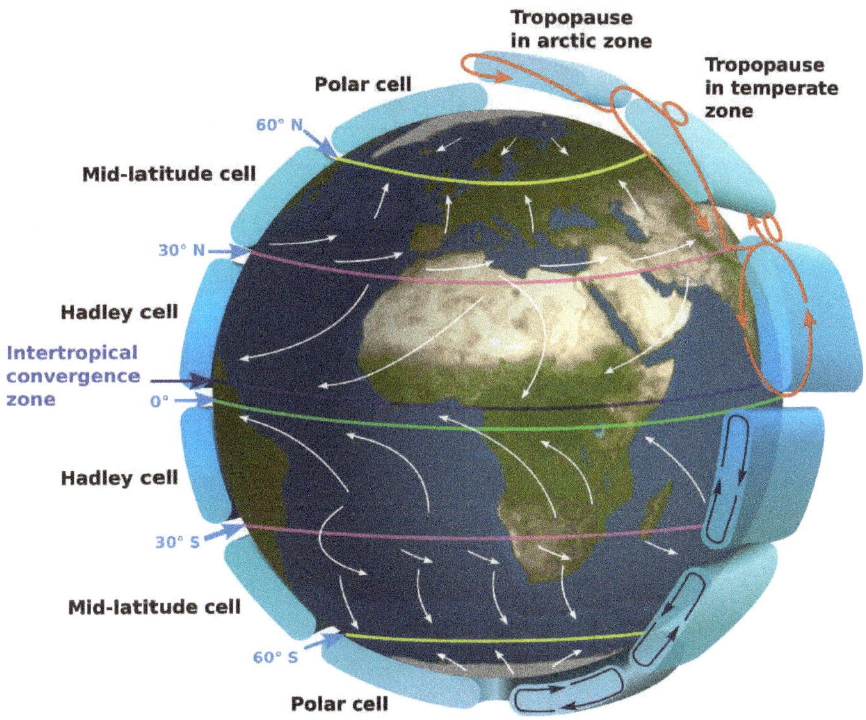

Idealised depiction (at equinox) of large scale atmospheric circulation on Earth.

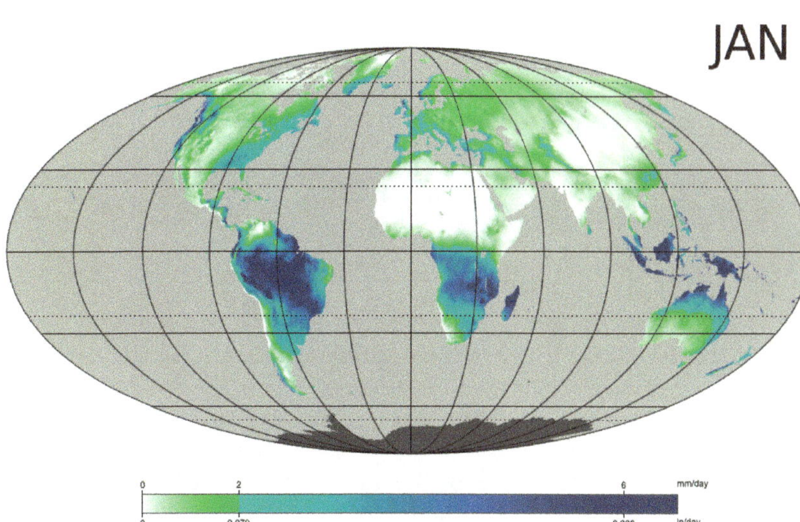

Long-term mean precipitation by month

Latitudinal Circulation Features

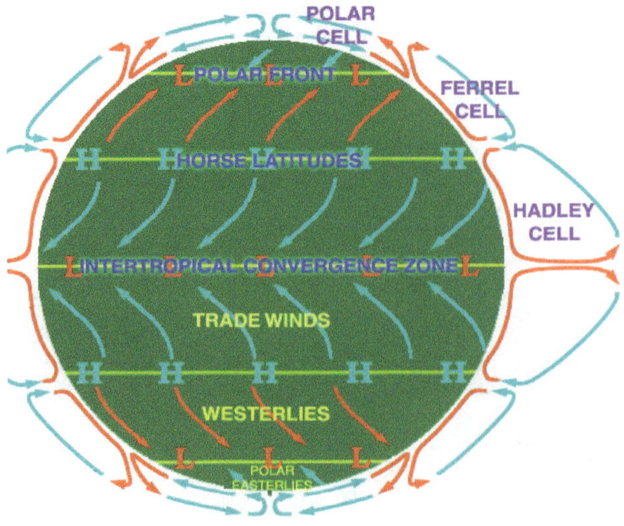

An idealised view of three large circulation cells showing surface winds

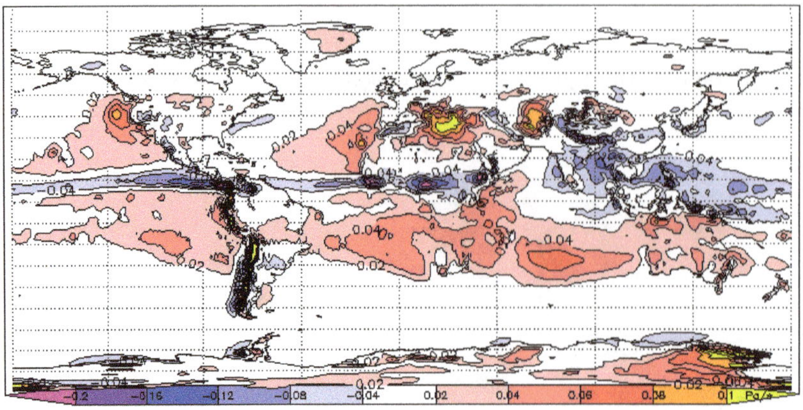

Vertical velocity at 500 hPa, July average. Ascent (negative values) is concentrated close to the solar equator; descent (positive values) is more diffuse but also occurs mainly in the Hadley cell.

The wind belts girdling the planet are organised into three cells; they are from the equator toward the poles, the Hadley cell, the Ferrel cell, and the Polar cell. Those cells exist in both the northern and southern hemispheres. The vast bulk of the atmospheric motion occurs in the Hadley cell. Note that there is one discrete Hadley cell that may split, shift and merge in a complicated process over time. The high pressure systems acting on the Earth's surface is balanced by the low pressure systems elsewhere. As a result, there a balance of forces acting on the Earth's surface.

Hadley Cell

The ITCZ's band of clouds over the Eastern Pacific and the Americas as seen from space

The atmospheric circulation pattern that George Hadley described was an attempt to explain the trade winds. The Hadley cell is a closed circulation loop, which begins at the equator. There, moist air is warmed by the Earth's surface, decreases in density and rises. A similar air mass rising on the other side of the equator forces those rising air masses to move poleward. The rising air creates a low pressure zone near the equator. As the air moves poleward, it cools, becomes more dense, and descends at about 30th parallel, creating a high-pressure area. The descended air then travels toward the equator along the surface, replacing the air that rose from the equatorial zone, closing the loop of the Hadley cell. The poleward movement of the air in the upper part of the troposphere deviates toward the east due to the coriolis acceleration (a manifestation of conservation of energy). At the ground level however, the movement of the air toward the equator in the lower troposphere deviates toward the west producing a wind from the east (again due to conservation of energy). The winds that flow to the west (from the east, easterly wind) at the ground level in the Hadley cell, are called the Trade Winds.

Though the Hadley cell is described as located at the equator, in the northern hemisphere, it shifts to higher latitudes June and July and toward lower latitudes December and January as it is caused by the Sun's heating of the surface. The zone where the greatest heating takes place is called the "thermal equator". As the southern hemisphere summer is December to March, the movement of the thermal equator to higher southern latitudes takes place then.

The Hadley system provides an example of a thermally direct circulation. The thermodynamic efficiency and power of the Hadley system, considered as a heat engine, is estimated as 2.6% and 200×10^{12} (TW).

Polar Cell

The **Polar cell**, likewise, is a simple system. Though cool and dry relative to equatorial air, the air masses at the 60th parallel are still sufficiently warm and moist to undergo convection and drive a thermal loop. At the 60th parallel, the air rises to the tropopause (about 8 km at this latitude) and moves poleward. As it does so, the upper level air mass deviates toward the east. When the air reaches the polar areas, it has cooled and is considerably denser than the underlying air. It descends, creating a cold, dry high-pressure area. At the polar surface level, the mass of air is driven toward the 60th parallel, replacing the air that rose there, and the Polar circulation cell is complete. As the air at the surface moves equatorward, it deviates toward the west. Again, the deviations of the air masses is due to conservation of energy, which is also referred to as the Coriolis effect. The air flows at the surface are called the Polar easterlies (easterly from the east).

The outflow of air mass from the cell creates harmonic waves in the atmosphere known as Rossby waves. These ultra-long waves determine the path of the jet stream, which travels within the transitional zone between the tropopause and the Ferrel cell. By acting as a heat sink, the Polar cell moves the abundant heat from the equator toward the polar regions.

The Hadley cell and the Polar cell are similar in that they are thermally direct; in other words, they exist as a direct consequence of surface temperatures. Their thermal characteristics drive the weather in their domain. The sheer volume of energy that the Hadley cell transports, and the depth of the heat sink that is the Polar cell, ensures that the effects of transient weather phenomena are not only not felt by the system as a whole, but — except under unusual circumstances — do not form. The endless chain of passing highs and lows which is part of everyday life for mid-latitude dwellers, at latitudes between 30 and 60° latitude, is unknown above the 60th and below the 30th parallels. There are some notable exceptions to this rule. In Europe, unstable weather extends to at least 70° north.

These atmospheric features are stable. Even though they may strengthen or weaken regionally over time, they do not vanish entirely.

The Polar cell, orography and Katabatic winds in Antarctica, can create very cold conditions at the surface, for instance the coldest temperature recorded on Earth: −89.2 °C at Vostok Station in Antarctica, measured 1983.

Ferrel Cell

Part of the air rising at 60° latitude diverges at high altitude toward the poles and creates the polar cell. The rest moves toward the equator where it collides at 30° latitude with the high level air of the Hadley cell. There it subsides and strengthens the high pressure ridges beneath. A large part of the energy that drives the Ferrel cell is provided by the Polar and Hadley cells circulating on either side and that drag the Ferrel cell with it. The Ferrel cell, theorized by William Ferrel (1817–1891), is therefore a secondary circulation feature, whose existence depends upon the Hadley and Polar cells on either side of it; it behaves much as an atmospheric ball bearing between the two. It might be thought of as an eddy created by the Hadley and Polar cells. The Ferrel cell is weak and the air flow and temperatures within are variable. For this reason, the mid latitudes are sometimes known as the **"zone of mixing."** At high altitudes, the Ferrel cell overrides the Hadley and Polar cells. The air of the Ferrel cell that descends at 30° latitude returns poleward at the ground level and as it does so it deviates toward the east. In the upper atmosphere of the Ferrel cell the air moving equatorward, deviates toward the west. Both of those deviations, as in the case of the Hadley and Polar cells, are driven by conservation of energy. As a result, just as the Easterly Trade Winds are found below the Hadley cell, the Westerlies are found beneath the Ferrel cell. The forces driving the flow in the Ferrel cell are weak and so the weather in that zone is variable. Thus, strong high-pressure areas which divert the prevailing westerlies, such as a Siberian high, can override the Ferrel cell, making it discontinuous.

While the Hadley and Polar cells are truly closed loops, the Ferrel cell is not, and the telling point is in the Westerlies, which are more formally known as the Prevailing Westerlies. The Easterly Trade Winds and the Polar Easterlies have nothing over which to prevail, as their parent circulation cells are strong enough, and face few obstacles either in the form of massive terrain features or high pressure zones. The weaker Westerlies of the Ferrel cell, however, can be disrupted. The local passage of a cold front may change that in a matter of minutes, and frequently does. As a result, at the surface, winds can vary abruptly in direction. But the winds above the surface, where they are less disrupted by terrain, are essentially westerly. A low pressure zone at 60° latitude that moves equator wards, or a high pressure zone at 30° latitude that moves poleward, will accelerate the Westerlies of the Ferrel cell. A strong high, moving polewards may bring Westerly winds for days. As the high and low pressure zones are established by the Hadley and Polar cells, it can be said that they drive the weather of the middle latitude Ferrel cell.

The Ferrel cell is driven by the Hadley and Polar cells. It has neither a strong source of heat nor a strong cold sink to drive convection. As a result, the weather within the Ferrel cell is highly variable and is influenced by changes to the Hadley and Polar cells. The base of the Ferrel cell is characterized by the movement of air masses, and the location of those air masses is influenced in part by the location of the jet stream, even though it flows near the tropopause. Overall, the movement of surface air is from the 30th latitude to the 60th. However, the upper flow of the Ferrel cell is weak and not well defined.

In contrast to the Hadley and Polar systems, the Ferrel system provides an example of a thermally indirect circulation. The Ferrel system acts as a heat pump with a coefficient of performance of 12.1, consuming kinetic energy at an approximate rate of 275 TW.

Longitudinal Circulation Features

While the Hadley, Ferrel, and Polar cells (whose axes are oriented along parallels i.e. latitudes) are the major features of global heat transport, they do not act alone. Temperature differences also drive a set of circulation cells, whose axis of circulation are longitudinally oriented. This atmospheric motion is known as zonal overturning circulation.

Latitudinal circulation is a result of the highest solar radiation per unit area (solar intensity) falling on the tropics. The solar intensity decreases as the latitude increases, reaching essentially zero at the poles. Longitudinal circulation, however, is a result of the heat capacity of water, it's absorptivity, and it's mixing. Water absorbs more heat than does the land, but its temperature does not rise as greatly as does the land. As a result, temperature variations on land are greater than on water. The Hadley, Ferrel, and Polar cells operate at the largest scale of thousands of kilometers (synoptic scale). But, even at mesoscales (a horizontal range of 5 to several hundred kilometres), this effect is noticeable. During the day, air warmed by the relatively hotter land rises, and as it does so it draws a cool breeze from the sea that replaces the risen air. At night, the relatively warmer water and cooler land reverses the process, and a breeze from the land, of air cooled by the land, is carried offshore by night. This described effect is daily (diurnal).

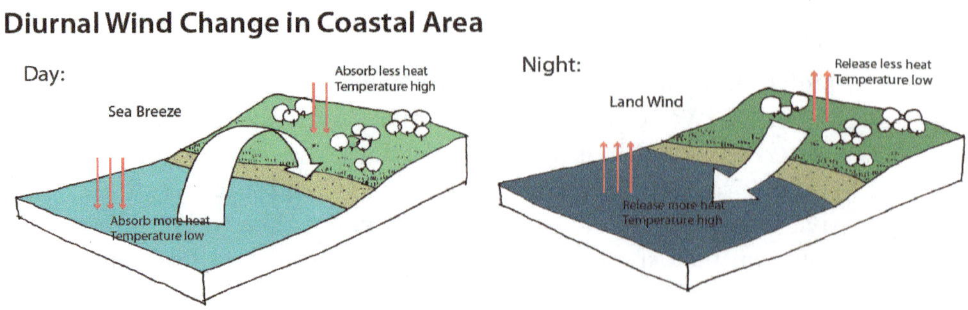

Diurnal wind change in local coastal area, also applies on the continental scale.

At the larger, synoptic, scale of oceans and continents, this effect is seasonal or even decadal. Warm air rising over the equatorial, continental, and western Pacific Ocean regions. When it reaches the tropopause, it cools and subsides in a region of relatively cooler water mass.

The Pacific Ocean cell plays a particularly important role in Earth's weather. This entirely ocean-based cell comes about as the result of a marked difference in the surface temperatures of the western and eastern Pacific. Under ordinary circumstances, the western Pacific waters are warm and the eastern waters are cool. The process begins when strong convective activity over equatorial East Asia and subsiding cool air off South America's west coast creates a wind pattern which pushes Pacific water westward and piles it up in the western Pacific. (Water levels in the western Pacific are about 60 cm higher than in the eastern Pacific, a difference due entirely to the force of moving air.)

Walker Circulation

The Pacific cell is of such importance that it has been named the Walker circulation after Sir Gilbert Walker, an early-20th-century director of British observatories in India, who sought a means of predicting when the monsoon winds of India would fail. While he was never successful in doing so, his work led him to the discovery of a link between the periodic pressure variations in the Indian Ocean, and those between the eastern and western Pacific, which he termed the "Southern Oscillation".

The movement of air in the Walker circulation affects the loops on either side. Under normal circumstances, the weather behaves as expected. But every few years, the winters become unusually warm or unusually cold, or the frequency of hurricanes increases or decreases, and the pattern sets in for an indeterminate period. The Walker Cell plays a key role in this and in the El Niño phenomenon. If convective activity slows in the Western Pacific for some reason (this reason is not currently known), the climate dominoes next to it begin to topple. First, the upper-level westerly winds fail. This cuts off the source of returning, cool air that would normally subside at about 30° north latitude, and therefore the air returning as surface Easterlies ceases.

The consequence of this is twofold. Warm water ceases to surge into the eastern Pacific from the west (it was piled by past easterly winds) since there is no longer a surface wind to push it into the area of the west pacific. This and the corresponding effects of the Southern Oscillation result in long-term unseasonable temperatures and precipitation patterns in North and South America, Australia, and Southeast Africa, and the disruption of ocean currents.

Meanwhile, in the Atlantic, fast-blowing upper level Westerlies of the Hadley cell, which would ordinarily be blocked by the Walker circulation and unable to reach such intensities, form. These winds tear apart the tops of nascent hurricanes and greatly diminish the number which are able to reach full strength.

El Niño – Southern Oscillation

El Niño and *La Niña* are opposite surface temperature anomalies of the Southern Pacific, which heavily influence the weather on a large scale. In the case of El Niño, warm surface water approaches the coasts of South America which results in blocking the upwelling of nutrient-rich deep water. This has serious impacts on the fish populations.

In the La Niña case, the convective cell over the western Pacific strengthens inordinately, resulting in colder than normal winters in North America, and a more robust cyclone season in South-East Asia and Eastern Australia. There is also an increased upwelling of deep cold ocean waters and more intense uprising of surface air near South America, resulting in increasing numbers of drought occurrences, although it is often argued that fishermen reap benefits from the more nutrient-filled eastern Pacific waters.

The neutral part of the cycle – the normal component – has been referred to humorously by some as La Nada, which means the nothing in Spanish.

Evolution of Earth's Atmosphere

Earliest Atmosphere

The first atmosphere would have consisted of gases in the solar nebula, primarily hydrogen. In addition, there would probably have been simple hydrides such as those now found in the gas giants (Jupiter and Saturn), notably water vapor, methane and ammonia. As the solar nebula dissipated, these gases would have escaped, partly driven off by the solar wind.

Second Atmosphere

Outgassing from volcanism, supplemented by gases produced during the late heavy bombardment of Earth by huge asteroids, produced the next atmosphere, consisting largely of nitrogen plus carbon dioxide and inert gases. A major part of carbon-dioxide emissions soon dissolved in water and built up carbonate sediments.

Researchers have found water-related sediments dating from as early as 3.8 billion years ago. About 3.4 billion years ago, nitrogen formed the major part of the then stable "second atmosphere". An influence of life has to be taken into account rather soon in the history of the atmosphere, because hints of early life-forms appear as early as 3.5 billion years ago.How Earth at that time maintained a climate warm enough for liquid water and life, if the early Sun put out 30% lower solar radiance than today, is a puzzle known as the "faint young Sun paradox".

The geological record however shows a continually relatively warm surface during the complete early temperature record of Earth - with the exception of one cold glacial phase about 2.4 billion years ago. In the late Archean Eon an oxygen-containing atmosphere began to develop, apparently produced by photosynthesizing cyanobacteria (see Great Oxygenation Event), which have been found as stromatolite fossils from 2.7 billion years ago. The early basic carbon isotopy (isotope ratio proportions) very much approximates current conditions, suggesting that the fundamental features of the carbon cycle became established as early as 4 billion years ago.

Ancient sediments in the Gabon dating from between about 2,150 and 2,080 million years ago provide a record of Earth's dynamic oxygenation evolution. These fluctuations in oxygenation were likely driven by the Lomagundi carbon isotope excursion.

Third Atmosphere

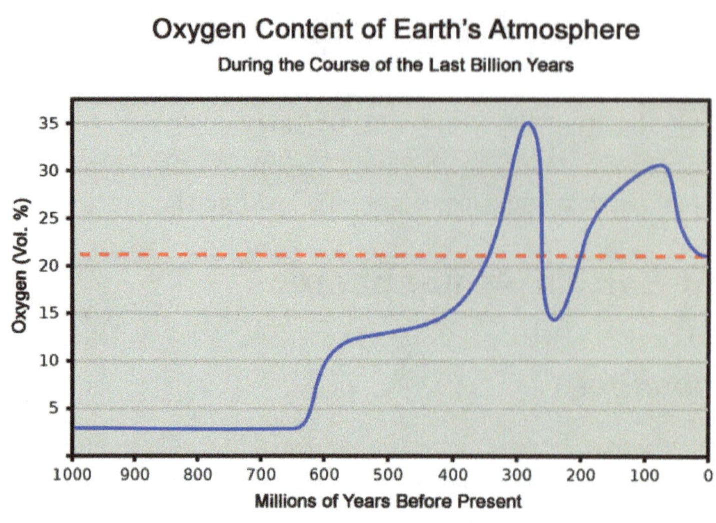

Oxygen content of the atmosphere over the last billion years. This diagram in more detail

The constant re-arrangement of continents by plate tectonics influences the long-term evolution of the atmosphere by transferring carbon dioxide to and from large continental carbonate stores. Free oxygen did not exist in the atmosphere until about 2.4 billion years ago during the Great Oxygenation Event and its appearance is indicated by the end of the banded iron formations. Before this time, any oxygen produced by photosynthesis was consumed by oxidation of reduced materials, notably iron. Molecules of free oxygen did not start to accumulate in the atmosphere until the rate of production of oxygen began to exceed the availability of reducing materials. This point signifies a shift from a reducing atmosphere to an oxidizing atmosphere. O_2 showed major variations until reaching a steady state of more than 15% by the end of the Precambrian. The following time span from 541 million years ago to the present day is the Phanerozoic Eon, during the earliest period of which, the Cambrian, oxygen-requiring metazoan life forms began to appear.

The amount of oxygen in the atmosphere has fluctuated over the last 600 million years, reaching a peak of about 30% around 280 million years ago, significantly higher than today's 21%. Two main processes govern changes in the atmosphere: Plants use carbon dioxide from the atmosphere, releasing oxygen. Breakdown of pyrite and volcanic eruptions release sulfur into the atmosphere, which oxidizes and hence reduces the amount of oxygen in the atmosphere. However, volcanic eruptions also release carbon dioxide, which plants can convert to oxygen. The exact cause of the variation of the amount of oxygen in the atmosphere is not known. Periods with much oxygen in the atmosphere are associated with rapid development of animals. Today's atmosphere contains 21% oxygen, which is high enough for this rapid development of animals.

This animation shows the buildup of tropospheric CO_2 in the Northern Hemisphere with a maximum around May. The maximum in the vegetation cycle follows, occurring in the late summer. Following the peak in vegetation, the drawdown of atmospheric CO_2 due to photosynthesis is apparent, particularly over the boreal forests.

The scientific consensus is that the anthropogenic greenhouse gases currently accumulating in the atmosphere are the main cause of global warming.

Images From Space

On October 19, 2015 NASA started a website containing daily images of the full sunlit side of Earth on http://epic.gsfc.nasa.gov/. The images are taken from the Deep Space Climate Observatory (DSCOVR) and show Earth as it rotates during a day.

Blue light is scattered more than other wavelengths by the gases in the atmosphere, giving Earth a blue halo when seen from space.

The geomagnetic storms cause beautiful displays of aurora across the atmosphere.

Limb view, of Earth's atmosphere. Colors roughly denote the layers of the atmosphere.

This image shows the Moon at the centre, with the limb of Earth near the bottom transitioning into the orange-colored troposphere. The troposphere ends abruptly at the tropopause, which appears in the image as the sharp boundary between the orange- and blue-colored atmosphere. The silvery-blue noctilucent clouds extend far above Earth's troposphere.

Earth's atmosphere backlit by the Sun in an eclipse observed from deep space onboard Apollo 12 in 1969.

Others

Other astronomical bodies such as these listed have known atmospheres.

In the Solar System

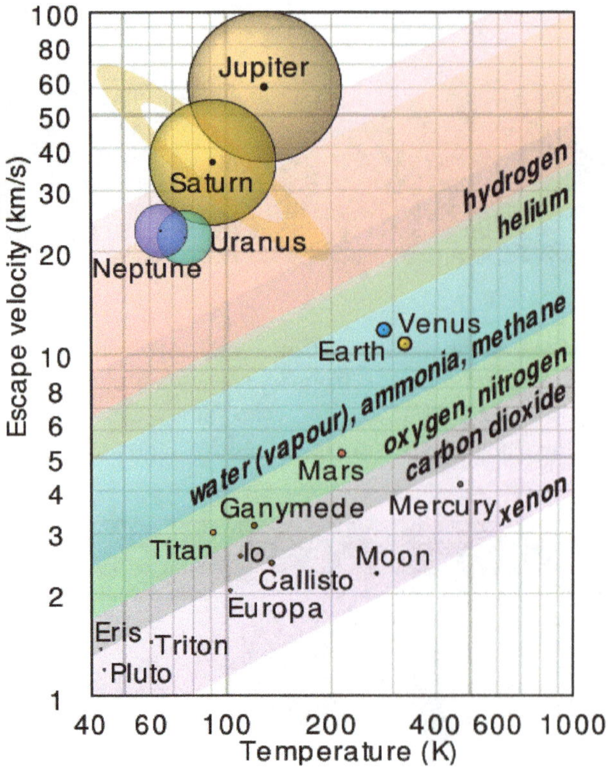

Graphs of escape velocity against surface temperature of some Solar System objects showing which gases are retained. The objects are drawn to scale, and their data points are at the black dots in the middle.

- Atmosphere of the Sun
- Atmosphere of Mercury
- Atmosphere of Venus
- Atmosphere of Earth
- Atmosphere of the Moon
- Atmosphere of Mars
- Atmosphere of Ceres
- Atmosphere of Jupiter
- Atmosphere of Io

- Atmosphere of Callisto
- Atmosphere of Europa
- Atmosphere of Ganymede
- Atmosphere of Saturn
- Atmosphere of Titan
- Atmosphere of Enceladus
- Atmosphere of Uranus
- Atmosphere of Titania
- Atmosphere of Neptune
- Atmosphere of Triton
- Atmosphere of Pluto

Outside the Solar System

- Atmosphere of HD 209458 b

Circulation

The circulation of the atmosphere occurs due to thermal differences when convection becomes a more efficient transporter of heat than thermal radiation. On planets where the primary heat source is solar radiation, excess heat in the tropics is transported to higher latitudes. When a planet generates a significant amount of heat internally, such as is the case for Jupiter, convection in the atmosphere can transport thermal energy from the higher temperature interior up to the surface.

Importance

From the perspective of a planetary geologist, the atmosphere is an evolutionary agent essential to the shaping of a planet. The wind picks up dust and other particles which when they collide with the terrain erodes the relief and leaves deposits (eolian processes). Frost and precipitations, which depend on the atmospheric composition, also influence the relief. Climate changes can influence a planet's geological history. Conversely, studying surface of Earth leads to an understanding of the atmosphere and climate of a planet — both its present state and its past.

For a meteorologist, the composition of the atmosphere determines the climate and its variations.

For a biologist or paleontologist, the atmospheric composition is closely dependent on the appearance of the life and its evolution.

REFERENCES

1. Henry George Liddell, Robert Scott, A Greek-English Lexicon, on Perseus Digital Library
2. σφαρα, Henry George Liddell, Robert Scott, A Greek-English Lexicon, on Perseus Digital Library
3. Seki, K.; Elphic, R. C.; Hirahara, M.; Terasawa, T.; Mukai, T. (2001). On Atmospheric Loss of Oxygen Ions from Earth Through Magnetospheric Processes. Science. 291 (5510): 1939–1941. Bibcode:2001Sci...291.1939S. doi:10.1126/science.1058913. PMID 11239148. Retrieved 2007-03-07.
4. Weaver, D.; Villard, R. (2007-01-31). Hubble Probes Layer-cake Structure of Alien World's Atmosphere. Hubble News Center. Retrieved 2007-03-11.
5. Zimmer, Carl (3 October 2013). Earth's Oxygen: A Mystery Easy to Take for Granted. New York Times. Retrieved 3 October 2013.
6. Lide, David R. Handbook of Chemistry and Physics. Boca Raton, FL: CRC, 1996: 14-7
7. Ultraviolet radiation in the solar system By Manuel Vázquez, Arnold Hanslmeier
8. allace, John M. and Peter V. Hobbs. Atmospheric Science; An Introductory Survey.Elsevier. Second Edition, 2006. ISBN 978-0-12-732951-2. Chapter 1
9. Trace Gases. Ace.mmu.ac.uk. Archived from the original on 9 October 2010. Retrieved 2010-10-16.
10. Source for figures: Carbon dioxide, NOAA Earth System Research Laboratory, (updated 2013-03). Methane, IPCCTAR table 6.1, (updated to 1998). The NASA total was 17 ppmv over 100%, and CO_2 was increased here by 15 ppmv. To normalize, N_2 should be reduced by about 25 ppmv and $O_2$by about 7 ppmv.
11. Vaughan, Adam (2015-05-06). Global carbon dioxide levels break 400ppm milestone. The Guardian. ISSN 0261-3077. Retrieved 2016-12-25.
12. http://www.nasa.gov/mission_pages/sunearth/science/mos-upper-atmosphere.html#.VHg5AzHF8vY
13. Randy Russell (2008). The Thermosphere. Retrieved 2013-10-18.
14. The height of the tropopause. Das.uwyo.edu. Retrieved 2012-04-18.
15. Ahrens, C. Donald. Essentials of Meteorology. Published by Thomson Brooks/Cole, 2005.
16. States, Robert J.; Gardner, Chester S. (January 2000). Thermal Structure of the Mesopause Region (80–105 km) at 40°N Latitude. Part I: Seasonal Variations. Journal of the Atmospheric Sciences 2000. 57: 66–77. Bibcode:2000JAtS...57...66S. doi:10.1175/1520-0469(2000)057<0066:TSOTMR>2.0.CO;2.
17. Joe Buchdahl. Atmosphere, Climate & Environment Information Programme. Ace.mmu.ac.uk. Archived from the original on 2010-07-01. Retrieved 2012-04-18.
18. Journal of the Atmospheric Sciences (1993). stratopause. Retrieved 2013-10-18.
19. Barry, R.G. & Chorley R.J. (1971). Atmosphere, Weather and Climate. London,

Menthuen & Co Ltd., p. 65

20. Tyson, P.D., Preston-Whyte, R.A. (2013). The Weather and Climate of Southern Africa. (Second Edition). Oxford, Oxford University Press. p. 4.
21. McGraw-Hill Concise Encyclopedia of Science & Technology. (1984). Troposphere. It contains about four-fifths of the mass of the whole atmosphere.
22. ISS022-E-062672 caption. NASA. Retrieved 21 September 2012.
23. ''homosphere''—AMS Glossary. Amsglossary.allenpress.com. Archived from the original on 14 September 2010. Retrieved 2010-10-16.
24. Earth's Atmosphere. Archived from the original on 2009-06-14.
25. NASA — Earth Fact Sheet. Nssdc.gsfc.nasa.gov. Archived from the original on 30 October 2010. Retrieved 2010-10-16.
26. Global Surface Temperature Anomalies.[permanent
27. Earth's Radiation Balance and Oceanic Heat Fluxes.
28. Coupled Model Intercomparison Project Control Run(PDF).
29. Geometric altitude vs. temperature, pressure, density, and the speed of sound derived from the 1962 U.S. Standard Atmosphere.
30. The Mass of the Atmosphere: A Constraint on Global Analyses. Ams.allenpress.com. 1970-01-01. Retrieved 2010-10-16.
31. Lutgens, Frederick K. and Edward J. Tarbuck (1995) The Atmosphere, Prentice Hall, 6th ed., pp14-17, ISBN 0-13-350612-6
32. Atmospheric Temperature Trends, 1979-2005 : Image of the Day. Earthobservatory.nasa.gov. 2000-01-01. Retrieved 2014-06-10.
33. Zahnle, K.; Schaefer, L.; Fegley, B. (2010). Earth's Earliest Atmospheres. Cold Spring Harbor Perspectives in Biology. 2 (10): a004895. doi:10.1101/cshperspect.a004895. PMC 2944365 . PMID 20573713.
34. B. Windley: The Evolving Continents. Wiley Press, New York 1984
35. J. Schopf: Earth's Earliest Biosphere: Its Origin and Evolution. Princeton University Press, Princeton, N.J., 1983
36. Timothy W. Lyons, Christopher T. Reinhard & Noah J. Planavsky (2014). Atmospheric oxygenation three billion years ago. Nature. 506 (7488): 307–15. Bibcode:2014Natur.506..307L. doi:10.1038/nature13068. PMID 24553238.
37. Christopher R. Scotese, Back to Earth History : Summary Chart for the Precambrian, Paleomar Project
38. Peter Ward: Out of Thin Air: Dinosaurs, Birds, and Earth's Ancient Atmosphere
39. Summary for Policymakers (PDF). Climate Change 2007: The Physical Science Basis. Contribution of Working Group I to the Fourth Assessment Report of the Intergovernmental Panel on Climate Change. Intergovernmental Panel on Climate Change. 5 February 2007.
40. Starting from Pollution—Definition from the Merriam-Webster Online Dictionary
41. Northon, Karen. Daily Views of Earth Available on New NASA Website. NASA.

Retrieved 2015-10-21.

42. Muriel Gargaud, Encyclopedia of Astrobiology, Volume 3, Springer Science & Business Media, May 26, 2011, p. 879.
43. David C. Catling and Kevin J. Zahnle, The Planetary Air Leak, Scientific American, May 2009, p. 26 (accessed 25 July 2012)
44. Shizgal, B. D.; Arkos, G. G. (1996). Nonthermal escape of the atmospheres of Venus, Earth, and Mars. Reviews of Geophysics. 34 (4): 483–505. Bibcode:1996RvGeo..34..483S. doi:10.1029/96RG02213.
45. Lammer, H.; Lichtenegger, H. I. M.; Biernat, H. K.; Erkaev, N. V.; Arshukova, I. L.; Kolb, C.; Gunell, H.; Lukyanov, A.; Holmstrom, M.; Barabash, S.; Zhang, T. L.; Baumjohann, W. (2006). Loss of hydrogen and oxygen from the upper atmosphere of Venus. Planetary and Space Science. 54 (13–14): 1445–1456. Bibcode:2006P&SS...54.1445L. doi:10.1016/j.pss.2006.04.022.
46. Webster, Guy (April 8, 2013). Remaining Martian Atmosphere Still Dynamic. NASA. Retrieved April 9, 2013.
47. Wall, Mike (April 8, 2013). Most of Mars' Atmosphere Is Lost in Space. Space.com. Retrieved April 9, 2013.
48. Lammer, H.; Stumptner, W.; Bauer, S. J. (1998). Dynamic escape of H from Titan as consequence of sputtering induced heating. Planetary and Space Science. 46 (9–10): 1207–1213. Bibcode:1998P&SS...46.1207L. doi:10.1016/S0032-0633(98)00050-6.
49. Wilson, J. K.; Mendillo, M.; Baumgardner, J.; Schneider, N. M.; Trauger, J. T.; Flynn, B. (2002). The dual sources of Io's sodium clouds. Icarus. 157 (2): 476–489. Bibcode:2002Icar..157..476W. doi:10.1006/icar.2002.6821.
50. Melosh, H.J.; Vickery, A.M. (April 1989). Impact erosion of the primordial atmosphere of Mars. Nature. 338: 487–489. Bibcode:1989Natur.338..487M. doi:10.1038/338487a0.
51. Space Studies Board, Division on Engineering and Physical Sciences (Jan 15, 1961). The Atmospheres of Mars and Venus. National Academies Press.

CHAPTER 2

Surface Weather Analysis

Chapter Contents

The surface weather map was the first weather map produced, dating back to the early 19th century. Even today, it remains the one of the most useful charts for ascertaining current weather conditions just above the surface of the earth for a large geographic region. These maps are called surface analysis charts if they contain fronts and analyzed pressure fields, with the solid lines representing isobars.

INTRODUCTION

Surface weather analysis is a special type of weather map that provides a view of weather elements over a geographical area at a specified time based on information from ground-based weather stations.

Weather maps are created by plotting or tracing the values of relevant quantities such as sea level pressure, temperature, and cloud cover onto a geographical map to help find synoptic scale features such as weather fronts.

The first weather maps in the 19th century were drawn well after the fact to help devise a theory on storm systems. After the advent of the telegraph, simultaneous surface weather observations became possible for the first time, and beginning in the late 1840s, the Smithsonian Institution became the first organization to draw real-time surface analyses. Use of surface analyses began first in the United States, spreading worldwide during the 1870s. Use of the Norwegian cyclone model for frontal analysis began in the late 1910s across Europe, with its use finally spreading to the United States during World War II.

Surface weather analyses have special symbols that show frontal systems, cloud cover, precipitation, or other important information. For example, an H may represent high pressure, implying clear skies and relatively warm weather. An L, on the other hand, may represent low pressure, which frequently accompanies precipitation. Various symbols are used not just for frontal zones and other surface boundaries on weather maps, but also to depict the present weather at various locations on the weather map. Areas of precipitation help determine the frontal type and location.

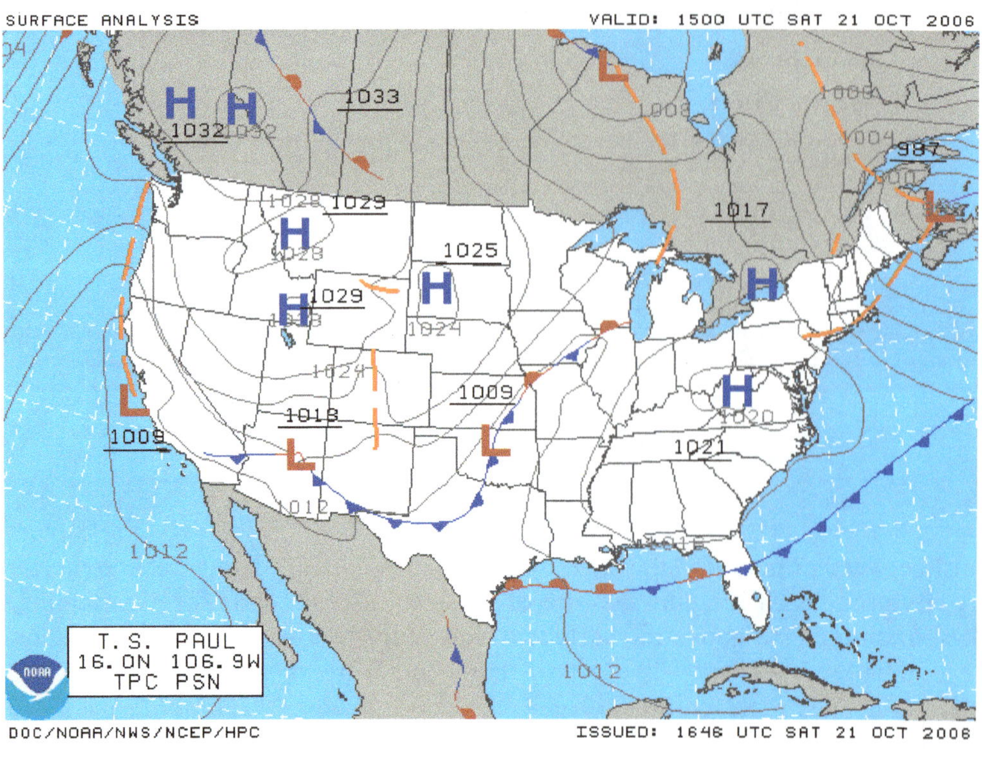

A surface weather analysis for the United States on October 21, 2006. By that time, Tropical Storm Paul was active (Paul later became a hurricane).

The use of weather charts in a modern sense began in the middle portion of the 19th century in order to devise a theory on storm systems. The development of a telegraph network by 1845 made it possible to gather weather information from multiple distant locations quickly enough to preserve its value for real-time applications. The Smithsonian Institution developed its network of observers over much of the central and eastern United States between the 1840s and 1860s. The U.S. Army Signal Corps inherited this network between 1870 and 1874 by an act of Congress, and expanded it to the west coast soon afterwards.

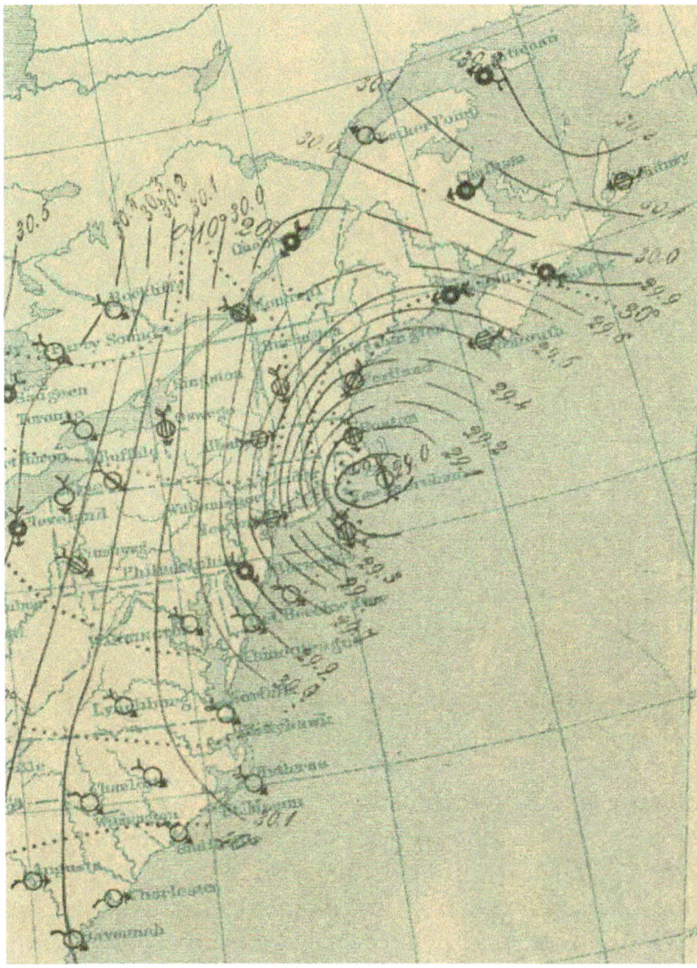

Surface analysis of Great Blizzard of 1888 on March 12, 1888 at 10 pm

The weather data was at first less useful as a result of the different times at which weather observations were made. The first attempts at time standardization took hold in Great Britain by 1855. The entire United States did not finally come under the influence of time zones until 1905, when Detroit finally established standard time. Other countries followed the lead of the United States in taking simultaneous weather observations, starting in 1873.Other countries then began preparing surface analyses. The use of frontal zones on weather maps did not appear until the introduction of the Norwegian cyclone model in the late 1910s, despite Loomis' earlier attempt at a similar notion in 1841. Since the leading edge of air mass changes bore resemblance to the military fronts of World War I, the term "front" came into use to represent these lines.

Present weather symbols used on weather maps

Despite the introduction of the Norwegian cyclone model just after World War I, the United States did not formally analyze fronts on surface analyses until late 1942, when the WBAN Analysis Center opened in downtown Washington, D.C.. The effort to automate map plotting began in the United States in 1969, with the process complete in the 1970s. Hong Kong completed their process of automated surface plotting by 1987. By 1999, computer systems and software had finally become sophisticated enough to allow for the ability to underlay on the same workstation satellite imagery, radar imagery, and model-derived fields such as atmospheric thickness and frontogenesis in combination with surface observations to make for the best possible surface analysis. In the United States, this development was achieved when Intergraph workstations were replaced by n-AWIPS workstations. By 2001, the various surface analyses done within the National Weather Service were combined into the Unified Surface Analysis, which is issued every six hours and combines the analyses of four different centers. Recent advances in both the fields of meteorology and geographic information systems have made it possible to devise finely tailored weather maps. Weather information can quickly be matched to relevant geographical detail. For instance, icing conditions can be mapped onto the road network. This will likely continue to lead to changes in the way surface analyses are created and displayed over the next several years. The pressureNET project is an ongoing attempt to gather surface pressure data using smartphones.

Station Model Used on Weather Maps

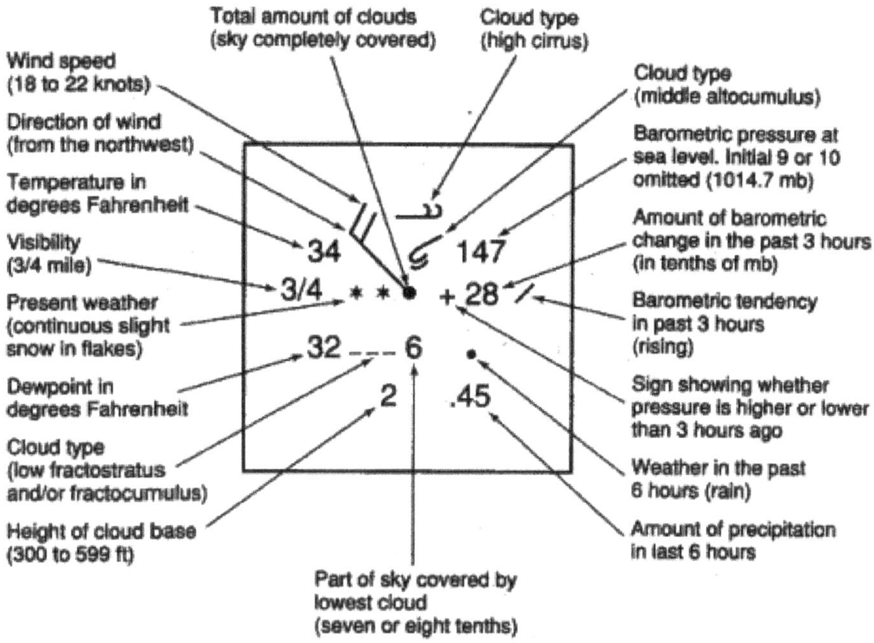

Station model plotted on surface weather analyses

When analyzing a weather map, a station model is plotted at each point of observation. Within the station model, the temperature, dewpoint, windspeed and direction, atmospheric pressure, pressure tendency, and ongoing weather are plotted. The circle in the middle represents cloud cover; fraction it is filled in represents the degree of overcast. Outside the United States, temperature and dewpoint are plotted in degrees Celsius. The wind barb points in the direction from which the wind is coming. Each full flag on the wind barb represents 10 knots (19 km/h) of wind, each half flag represents 5 knots (9 km/h). When winds reach 50 knots (93 km/h), a filled in triangle is used for each 50 knots (93 km/h) of wind. In the United States, rainfall plotted in the corner of the station model are in inches. The international standard rainfall measurement unit is the millimeter. Once a map has a field of station models plotted, the analyzing isobars (lines of equal pressure), isallobars (lines of equal pressure change), isotherms (lines of equal temperature), and isotachs (lines of equal wind speed) are drawn. The abstract weather symbols were devised to take up the least room possible on weather maps.

Synoptic Scale Features

A synoptic scale feature is one whose dimensions are large in scale, more than several hundred kilometers in length. Migratory pressure systems and frontal zones exist on this scale.

Pressure Centers

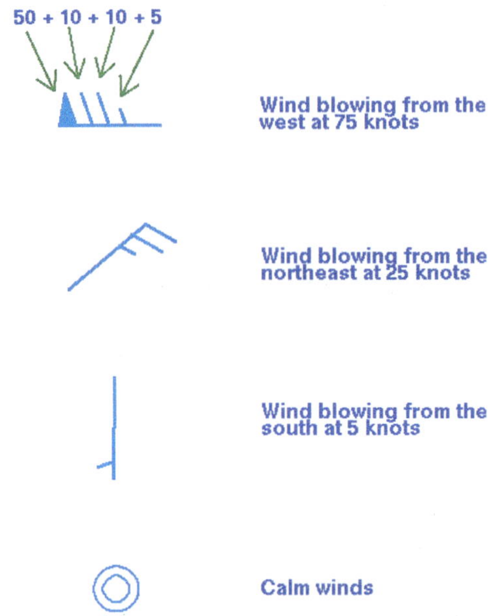

Wind barb interpretation

Centers of surface high- and low-pressure areas are found within closed isobars on a surface weather analysis are the absolute maxima and minima in the pressure field, and can tell a user in a glance what the general weather is in their vicinity. Weather maps in English-speaking countries will depict their highs as Hs and lows as Ls, while Spanish-speaking countries will depict their highs as As and lows as Bs.

Low Pressure

Low-pressure systems, also known as cyclones, are located in minima in the pressure field. Rotation is inward at the surface and counterclockwise in the northern hemisphere as opposed to inward and clockwise in the southern hemisphere due to the Coriolis force. Weather is normally unsettled in the vicinity of a cyclone, with increased cloudiness, increased winds, increased temperatures, and upward motion in the atmosphere, which leads to an increased chance of precipitation. Polar lows can form over relatively mild ocean waters when cold air sweeps in from the ice cap. The relatively warmer water leads to upward convection, causing a low to form, and precipitation usually in the form of snow. Tropical cyclones and winter storms are intense varieties of low pressure. Over land, thermal lows are indicative of hot weather during the summer.

High Pressure

High-pressure systems, also known as anticyclones, rotate outward at the surface and clockwise in the northern hemisphere as opposed to outward and counterclockwise in the southern hemisphere. Under surface highs, sinking of the atmosphere slightly warms the air by compression, leading to clearer skies, winds that are lighter, and a reduced chance of precipitation. The descending air is dry, hence less energy is required to raise its temperature. If high pressure persists, air pollution will build up due to pollutants trapped near the surface caused by the subsiding motion associated with the high.

Fronts

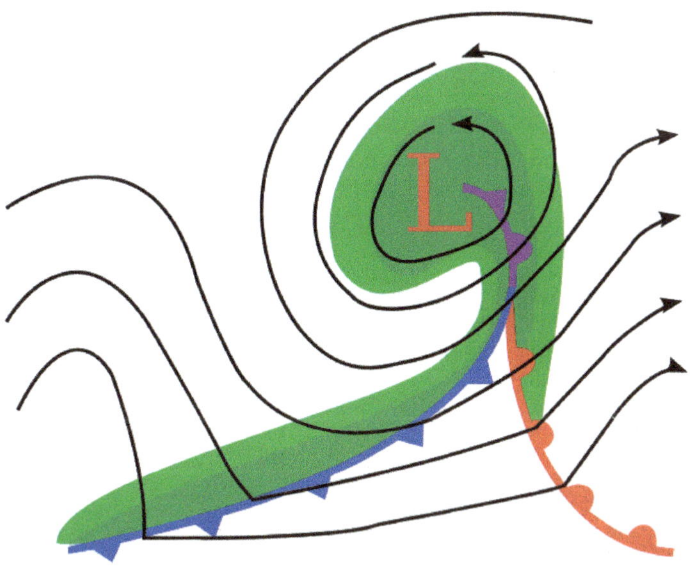

Occluded cyclone example. The triple point is the intersection of the cold, warm, and occluded fronts.

Fronts in meteorology are the leading edges of air masses that have a density, air temperature, and humidity different from the air mass it is invading. When an air mass passes over an area, it is marked by changes in temperature, moisture, wind speed and direction, atmospheric pressure, and often a change in the precipitation pattern. The change is called a front, whether it be warm or cold. Cold fronts develop when cold air masses, usually moving equatorward from the polar region high pressure zones, interact with warm moist air of low pressure systems. The fronts develop at the leading edge of that cold air mass and wrap around (counter-clockwise in the northern hemisphere) about the low pressure zone. A polar front may form approximately along the equatorward edge of the high-level polar jet. Fronts are guided by winds aloft, but they normally move at lesser speeds than those winds. In the northern hemisphere, they usually travel from west to east, although they can move in a north-south direction as well as they may wrap around an associated low pressure zone. Movement is driven by the pressure gradient force (horizontal differences in atmospheric pressure) and the Coriolis force (caused by the Earth's rotation on its axis). Frontal zones can be distorted by such geographic features as mountains and large bodies of water.

COLD FRONT

A cold front is defined as the leading edge of a cooler mass of air, replacing at ground level a warmer mass of air, which lies within a fairly sharp surface trough of low pressure. It forms in the wake of an extratropical cyclone, at the leading edge of its cold air advection pattern, which is also known as the cyclone's dry conveyor belt circulation. Temperature changes across the boundary can exceed 30 °C (54 °F). When enough moisture is present, rain can occur along the boundary. If there is significant instability along the boundary, a narrow line of thunderstorms can form along the frontal zone. If instability is less, a broad shield of rain can move in behind the front, which increases the temperature difference across the boundary. Cold fronts are stronger in the fall and spring transition seasons and weakest during the summer. When a cold front catches up with the preceding warm front, the portion of the boundary that does so is then known as an occluded front.

The symbol of a cold front: a blue line with triangles pointing in the direction of travel

A retreating cold front with clear air behind

A cold front over the eastern and central region of the United States of America

Development of Cold Fronts

The cooler and denser air wedges under the less-dense warmer air, lifting it. This upward motion causes lowered pressure along the cold front and can cause the formation of a narrow line of showers and thunderstorms when enough moisture is present. On weather maps, the surface position of the cold front is marked with the symbol of a blue line of triangles/spikes (pips) pointing in the direction of travel. A cold front's location is at the leading edge of the temperature drop off, which in an isotherm analysis would show up as the leading edge of the isotherm gradient, and it normally lies within a sharp surface trough. Cold fronts move faster than warm fronts and can produce sharper changes in weather. Since cold air is denser than warm air, it rapidly replaces the warm air preceding the boundary.

In the northern hemisphere, a cold front usually causes a shift of wind from southwest to northwest clockwise, also known as veering, and in the southern hemisphere a shift from northwest to southwest (anticlockwise, backing). Normally, cold fronts can be marked by these characteristics:

Weather phenomenon	Prior to the passing of the front	While the front is passing	After the passing of the front
Temperature	Warm	Cooling suddenly	Steadily cooling
Atmospheric pressure	Decreasing steadily	Lowest, then sudden increase	Increasing steadily

Winds	• Southwest to southeast (northern hemisphere) • Northwest to northeast (southern hemisphere)	Gusty; shifting	• North to west, usually northwest (northern hemisphere) • South to west, usually southwest (southern hemisphere)
Precipitation/conditions*	Light patchy rain can be produced by stratocumulus or stratus in the warm sector. In summer, sometimes thunderstorms if a preceding squall line is present.	Prolonged rain (nimbostratus) or thunderstorms (cumulonimbus): depends on conditions.	Showers, then clearing
Clouds*	Often preceded by cirrus, cirrostratus then altostratus like a warm front (but usually with smaller amounts of these clouds). Areas of cirrocumulus and altocumulus within cirrostratus and altostratus more commonly seen than at a warm front. Larger cumulus clouds under the higher cloud types than at a warm front, where stratocumulus and cumulus humilis usually occur. Some of these cumulus clouds may produce showers ahead of the front.	Cumulonimbus and cumulus congestus producing frequent showers, with a sheet of upper altostratus, through which the sun can sometimes be seen. Less commonly nimbostratus occurs with continuous rain.	Patchy altocumulus or stratocumulus and higher cirrus clouds along with fast moving stratus fractus then eventually scattered cumulus and sometimes cumulonimbus.
Visibility*	Fair to poor in haze	Poor, but improving	Good, except in showers
Dew Point	High, steady	Sudden drop	Falling

*provided there is sufficient moisture.

Clouds

If the cold front is highly unstable, cumulonimbus clouds producing thunderstorms commonly form along the front. Anvil cirrus clouds may spread a considerable distance downwind from the thunderstorms. The other cloud types associated with a cold front depend on atmospheric conditions such as air mass stability and wind shear. As the front approaches, middle-étage gives way to altostratus and low-level stratocumulus with intermittent light precipitation if the warm airmass being displaced by the cold front is mostly stable. With significant airmass instability, vertically developed cumulus or cumulonimbus with showers and thunderstorms will form along the front.

After the passage of the cold front, the sky usually clears as high pressure builds in behind the system, although significant amounts of cumulus or stratocumulus, often in the form of long bands called *cloud streets* may persist if the air mass behind the front remains humid. Small and unchanging amounts of cumulus or cirrus clouds in an otherwise clear sky are usually indications of continuing fair weather as long as the barometric pressure remains comparatively high.

Precipitation

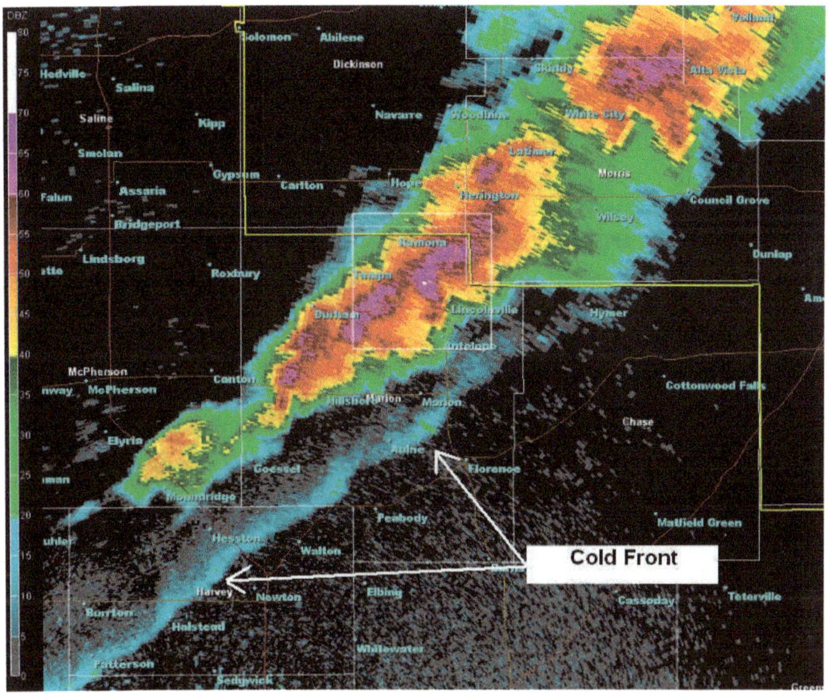

A cold front as it appeared on the National Weather Service Wichita, Kansas WSR-88D on April 3, 2011. The thin blue line labeled "cold front" is the front, with severe thunderstorms seen developing behind the front, which is moving towards the bottom right.

A cold front commonly brings a narrow band of precipitation that follows along the leading edge of the cold front. These bands of precipitation are often very strong in nature, and can bring severe thunderstorms, hailstorms and/or tornadoes. In the spring, these cold fronts can be very strong, and can bring strong winds when the pressure gradient is tighter than normal. During the winter months, cold fronts sometimes come through an area with little or no precipitation. Wider rain bands can occur behind cold fronts which tend to have more stratiform, and less convective, precipitation. These rainstorms sometimes bring flooding, and can move very slowly when the storm steering it is strong and embedded within a meridional flow pattern (with more pole to equator motion rather than west to east motion). In the winter, cold fronts can bring cold spells, and occasionally snow. In the spring or summer in temperate latitudes, hail may occasionally fall along with the rain. If moisture is not sufficient, such as when a system has previously moved across a mountain barrier, cold fronts can pass without cloudiness.

Undercutting

The idea that cold air wedges, or undercuts, the warm air is often used to depict how advancing cold fronts force warm air to rise along the sloping cold air, much like a snow shovel scooping up snow, creating instability. Provided with sufficient moisture, the rising air would thus condense, creating storms, clouds, and/or rain. While this concept is used to generally describe frontal precipitation patterns, it is technically incorrect.

Frontogenetical Circulation

Frontogenesis is the process of creating or steepening the temperature gradient of a front. During this process the atmosphere reacts in an attempt to restore balance, the consequence is a circular motion along the front where air is being lifted up, along the cold front and dropping downward, behind the frontal boundary. This is the actual force of upward motion along a front that is responsible for clouds and precipitation.

As the temperature gradient steepens during frontogenesis, the thermal wind becomes imbalanced. To maintain balance, the geostrophic wind aloft and below adjust, such that regions of divergence/convergence form. Mass continuity would require a vertical transport of air along the cold front where there is divergence (lowered pressure). Although this circulation is described by a series of processes, they are actually occurring at the same time, observable along the front as a thermally direct circulation. There are several factors that influence the final shape and tilt of the circulation around the front, ultimately determining the kind and location of clouds and precipitation.

Temperature Changes

Cold fronts are the leading edge of cooler air masses, hence the name "cold front". They have stronger temperature changes during the fall (autumn) and spring and during the middle of winter. Temperature changes associated with cold fronts can be as much as 50 °F (30 °C). When cold fronts come through, there is usually a quick, yet strong gust of wind, that shows that the cold front is passing. In surface weather observations, a remark known as FROPA is coded when this occurs. The effects from a cold front can last from hours to days. The air behind the front is cooler than the air it is replacing and the warm air is forced to rise, so it cools. As the cooler air cannot hold as much moisture as warm air, clouds form and rain occurs.

WARM FRONT

A warm front is a density discontinuity located at the leading edge of a homogeneous warm air mass, and is typically located on the equator-facing edge of an isotherm gradient. Warm fronts lie within broader troughs of low pressure than cold fronts, and move more slowly than the cold fronts which usually follow because cold air is denser and less easy to remove from the Earth's surface. This also forces temperature differences across warm fronts to be broader in scale. Clouds ahead of the warm front are mostly stratiform, and rainfall gradually increases as the front approaches. Fog can also occur preceding a warm frontal passage. Clearing and warming is usually rapid after frontal passage. If the warm air mass is unstable, thunderstorms may be embedded among the stratiform clouds ahead of the front, and after frontal passage thundershowers may continue. On weather maps, the surface location of a warm front is marked with a red line of semicircles pointing in the direction of travel.

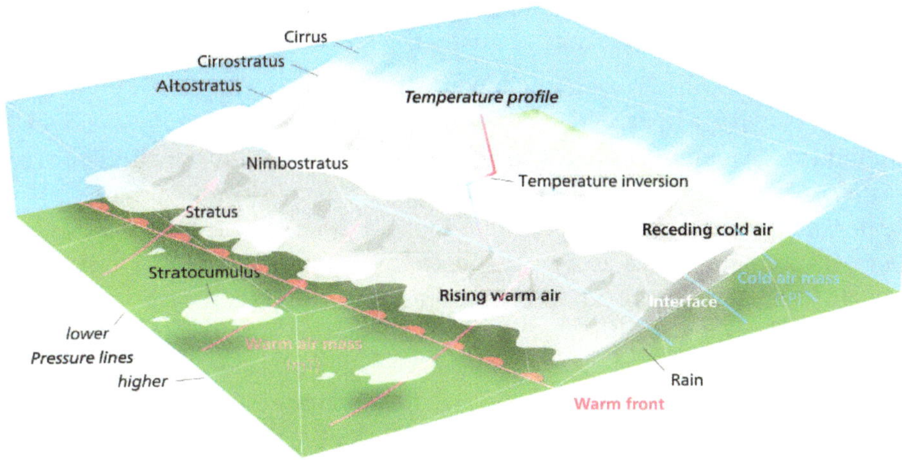

Illustration of a warm front. The warm air behind the front is slowly overtaking the cold air ahead of the front, which is moving more slowly in the same direction. The warmer air, due to lower density, climbs over the colder air as it moves. As a result of its increased altitude, it cools off and its moisture condenses, forming clouds and possibly precipitation.

Development

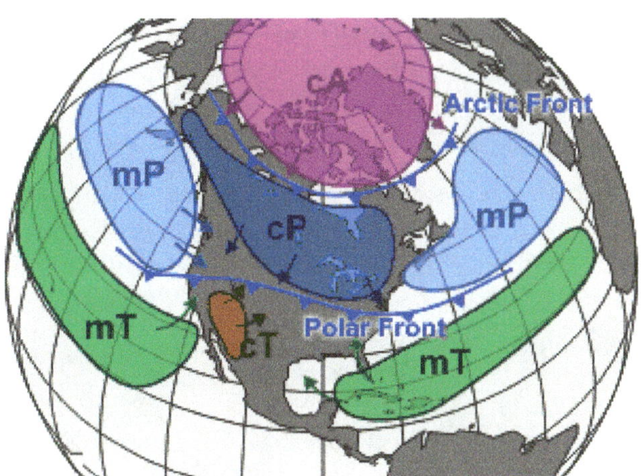

Different air masses that affect North America, as well as other continents, tend to be separated by frontal boundaries.

Air masses are large bodies of air with similar properties of temperature and humidity that form over source regions. The warm air mass behind a warm front is not only warmer, but often (but not always) also higher in humidity than the colder air preceding it. Because of a warm air mass's higher temperature and thus lesser density, mixing between the two air masses is unlikely. Being light, the warm air mass is unable to displace the cooler air mass and instead is forced upward along the upper boundary of the colder air in a process known as overrunning. The boundary between the two air masses has a gradual slope of 1:200 and lifting is slow but persistent.

As the air mass rises into regions of lower pressure, it expands and cools. As it cools, any water vapor that is present will condense and form extensive cloud cover. The first clouds that indicate an approaching warm front tend to be mostly high cirrus at first, changing to cirrostratus as the front approaches. However, if cirrocumulus also appears, there is greater airmass instability approaching ahead of the front. When these high clouds progressively invade the sky and the barometric pressure begins to fall, precipitation associated with the disturbance is likely about 6 to 8 hours away. A thickening and lowering of these high clouds into middle-étage altostratus or altocumulus is a good sign the warm front or low has moved closer and precipitation may begin within less than six hours. Once the clouds have thickened to 2,500 metres (8,200 ft) from the earth's surface, precipitation can begin to fall from heavy nimbostratus. If unstable altocumulus castellanus accompanies or takes the place of the main altostratus layer, cumulus congestus or cumulonimbus producing showers or thunderstorms may follow. Low stratus and stratocumulus commonly form underneath the main precipitating clouds.

A warm front is also defined as the transition zone where a warmer air mass is replacing a cooler air mass. Warm fronts generally move from southwest to northeast. If the warmer air originates over the ocean, it is not only warmer but also more moist than the air ahead of it.

Characteristics

If the air mass is relatively stable, rainfall will increase until the front reaches the location, at which time the clouds can extend all the way to the earth's surface as fog. Once the front passes, the location experiences some warming and clearing. If the air mass is unstable, thunderstorms may precede and follow the front and temperature changes will be larger.

In the northern hemisphere, a warm front causes a shift of wind blowing from southeast to southwest, and in the southern hemisphere a shift from winds blowing from northeast to northwest. Common characteristics associated with warm fronts include:

Weather phenomenon	Prior to the Passing of the Front	While the Front is Passing	After the Passing of the Front
Temperature	Cool	Warming suddenly	Warmer, then leveling off
Atmospheric pressure	Decreasing steadily	Leveling off	Slight rise followed by a decrease

Winds	• South to southeast (backing) (northern hemisphere) • North to northeast (veering) (southern hemisphere)	Variable	• South to southwest (veering) (northern hemisphere) • North to northwest (backing) (southern hemisphere)
Precipitation	Usually none, but in summer or warm temperatures, cumulus congestus may continue to exist under cirrostratus and altostratus creating light to moderate showers.	Persistent rain, usually moderate with some lighter periods and some heavier bursts.	Light drizzle, gradually ceasing.
Clouds	Cirrus, cirrostratus, altostratus, nimbostratus, then stratus (pilots use the acronym CCANS). Other clouds can also often be seen, including cirrocumulus amongst the approaching cirrus, altocumulus with or instead of altostratus (particularly if the front is weak), and occasionally cumulonimbus along with or instead of nimbostratus in summer. Additionally, stratocumulus often appears underneath the main altostratus deck and stratus fractus typically forms in precipitation falling from the thick nimbostratus layer. Often in warm temperatures, rain bearing cumulus congestus clouds can appear under the cirrostratus, and more rarely altocumulus castellanus clouds if convection is sufficient. In cold humid conditions, low airmass stratus or fog may obscure the main frontal clouds.	Nimbostratus, sometimes cumulonimbus	Clearing with scattered stratus and stratocumulus. If the warm front is part of a depression, there is often a sheet of altostratus (often broken in places to altocumulus) above this which thickens when the cold front approaches.
Visibility	Poor	Poor, but improving	Sunny
Dew point	Steady rise	Steady	Rise, then steady

Warm Sector

The warm sector is the area of warmer air behind a warm front, usually between the warm and cold fronts in a depression. Temperatures are often warmer than they are before the warm front or after the cold front. Cloud types can be mixed, but usually consist of stratocumulus, which can range to being broken to covering the entire sky depending on distance from the centre of low pressure. Temperature rises and the dew point remains steady. Wind direction and speed remains steady.

Depiction

On weather maps, the surface location of a warm front is marked with a red line of half circles pointing in the direction of the front. On colored weather maps, warm fronts are illustrated with a solid red line.

Occluded Front

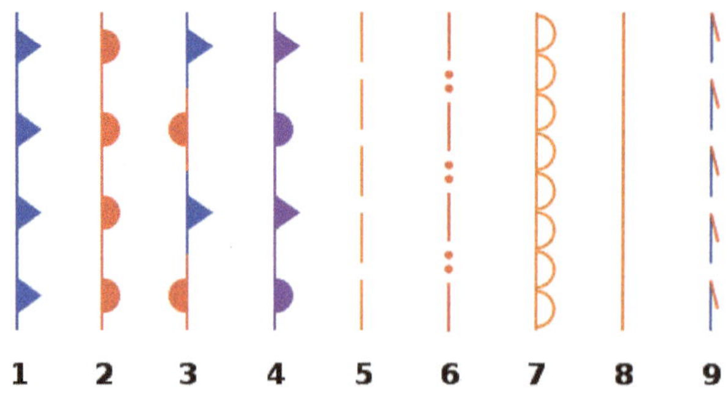

A guide to the symbols for weather fronts that may be found on a weather map:

- cold front
- warm front
- stationary front
- occluded front
- surface trough
- squall line
- dry line
- tropical wave
- Trowal

An occluded front is formed during the process of cyclogenesis when a cold front overtakes a warm front. The cold and warm fronts curve naturally poleward into the point of occlusion, which is also known as the triple point in meteorology. It lies within a sharp trough, but the air mass behind the boundary can be either warm or cold. In a cold occlusion, the air mass overtaking the warm front is cooler than the cool air ahead of the warm front, and plows under both air masses. In a warm occlusion, the air mass overtaking the warm front is not as cool as the cold air ahead of the warm front, and rides over the colder air mass while lifting the warm air. A wide variety of weather can be found along an occluded front, with thunderstorms possible, but usually their passage is associated with a drying of the air mass. Occluded fronts are indicated on a weather map by a purple line with alternating half-circles and triangles pointing in direction of travel. Occluded fronts usually form around mature low pressure areas.

The trowal is the projection on the Earth's surface of a tongue of warm air aloft, such as may be formed during the occlusion process of a depression.

Stationary Fronts and Shearlines

A stationary front is a non-moving boundary between two different air masses, neither of which is strong enough to replace the other. They tend to remain in the same area for long periods of time, usually moving in waves. There is normally a broad temperature gradient behind the boundary with more widely spaced isotherms. A wide variety of weather can be found along a stationary front, but usually clouds and prolonged precipitation are found there. Stationary fronts will either dissipate after several days or devolve into shear lines, but can change into a cold or warm front if conditions aloft change causing a driving of one air mass or the other. Stationary fronts are marked on weather maps with alternating red half-circles and blue spikes pointing in opposite directions, indicating no significant movement.

When stationary fronts become smaller in scale, degenerating to a narrow zone where wind direction changes over a short distance, they become known as shear lines. If the shear line becomes active with thunderstorms, it may support formation of a tropical storm or a regeneration of the feature back into a stationary front. A shear line is depicted as a line of red dots and dashes.

Mesoscale Features

Mesoscale features are smaller than synoptic scale systems like fronts, but larger than storm-scale systems like thunderstorms. Horizontal dimensions generally range from over ten kilometres to several hundred kilometres.

Dry Line

The dry line is the boundary between dry and moist air masses east of mountain ranges with similar orientation to the Rockies, depicted at the leading edge of the dew point, or moisture, gradient. Near the surface, warm moist air that is denser than warmer, dryer air wedges under the drier air in a manner similar to that of a cold front wedging under warmer air. When the warm moist air wedged under the drier mass heats up, it becomes less dense and rises and sometimes forms thunderstorms. At higher altitudes, the warm moist air is less dense than the cooler, drier air and the boundary slope reverses. In the vicinity of the reversal aloft, severe weather is possible, especially when a triple point is formed with a cold front.

During daylight hours, drier air from aloft drifts down to the surface, causing an apparent movement of the dryline eastward. At night, the boundary reverts to the west as there is no longer any solar heating to help mix the lower atmosphere. If enough moisture converges upon the dryline, it can be the focus of afternoon and evening thunderstorms. A dry line is depicted on United States surface analyses as a brown line with scallops, or bumps, facing into the moist sector. Dry lines are one of the few surface fronts where the special shapes along the drawn boundary do not necessarily reflect the boundary's direction of motion.

Outflow Boundaries and Squall Lines

A shelf cloud such as this one can be a sign that a squall is imminent

Organized areas of thunderstorm activity not only reinforce pre-existing frontal zones, but they can outrun cold fronts. This outrunning occurs in a pattern where the upper level jet splits into two streams. The resultant mesoscale convective system (MCS) forms at the point of the upper level split in the wind pattern in the area of best low-level inflow. The convection then moves east and equatorward into the warm sector, parallel to low-level thickness lines. When the convection is strong and linear or curved, the MCS is called a squall line, with the feature placed at the leading edge of the significant wind shift and pressure rise. Even weaker and less organized areas of thunderstorms will lead to locally cooler air and higher pressures, and outflow boundaries exist ahead of this type of activity, "SQLN" or "SQUALL LINE", while outflow boundaries are depicted as troughs with a label of "OUTFLOW BOUNDARY" or "OUTFLOW BNDRY".

Sea and Land Breeze Fronts

Sea breeze fronts occur on sunny days when the landmass warms the air above it to a temperature above the water temperature. Similar boundaries from downwind on lakes and rivers during the day, as well as offshore landmasses at night. Since the specific heat of water is so high, there is little diurnal temperature change in bodies of water, even on the sunniest days. The water temperature varies less than 1 °C (1 to 2 °F). By contrast, the land, with a lower specific heat, can vary several degrees in a matter of hours.

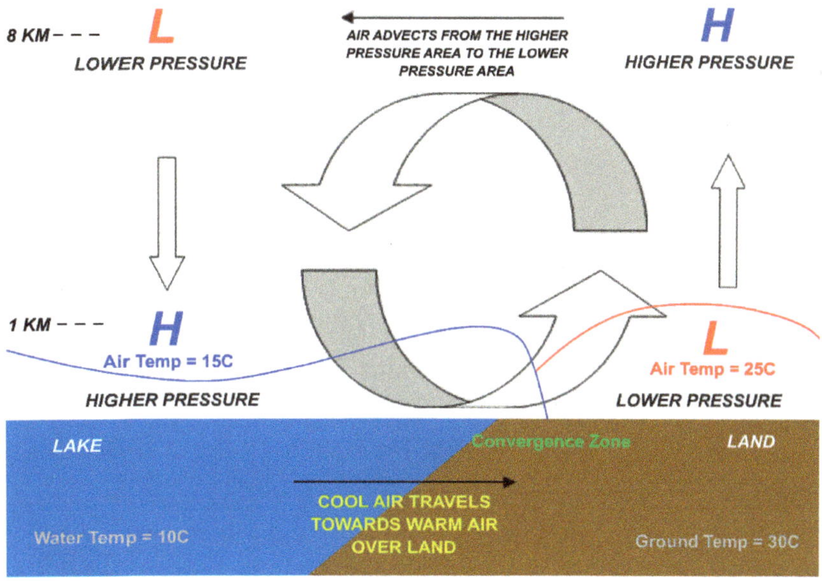

Idealized circulation pattern associated with a sea breeze

During the afternoon, air pressure decreases over the land as the warmer air rises. The relatively cooler air over the sea rushes in to replace it. The result is a relatively cool onshore wind. This process usually reverses at night where the water temperature is higher relative to the landmass, leading to an offshore land breeze. However, if water temperatures are colder than the land at night, the sea breeze may continue, only somewhat abated. This is typically the case along the California coast, for example.

If enough moisture exists, thunderstorms can form along sea breeze fronts that then can send out outflow boundaries. This causes chaotic wind/pressure regimes if the steering flow is weak. Like all other surface features, sea breeze fronts lie inside troughs of low pressure.

REFERENCES

1. Air Apparent: How Meteorologists Learned to Map, Predict, and Dramatize Weather. University of Chicago PressChicago: 1999.
2. Eric R. Miller. American Pioneers in Meteorology.Retrieved on 2007-04-18.
3. Human Intelligence.Francis Galton. Retrieved on 2007-04-18.
4. Frank Rives Millikan. Smithsonian Institution. Joseph Henry: Father of the Weather Service. Retrieved on 2006-10-22. Archived October 20, 2006, at the Wayback Machine.
5. WebExhibits. Daylight Saving Time. Retrieved on 2007-06-24.
6. NOAA. An Expanding Presence. Retrieved on 2007-05-05.

7. David M. Schultz. Perspectives on Fred Sanders's Research on Cold Fronts, 2003, revised, 2004, 2006, p. 5. Retrieved on 2006-07-14.
8. Bureau of Meteorology. Air Masses and Weather Maps.Retrieved on 2006-10-22.
9. Hydrometeorological Prediction Center. A Brief History of the Hydrometeorological Prediction Center. Retrieved on 2007-05-05.
10. ESSA. Prospectus for an NMC Digital Facsimile Incoder Mapping Program. Retrieved on 2007-05-05.
11. Hong Kong Observatory. The Hong Kong Observatory Computer System and Its Applications. Retrieved on 2007-05-05.
12. Hydrometeorological Prediction Center. Hydrometeorological Prediction Center 1999 Accomplishment Report. Retrieved on 2007-05-05.
13. David Roth. Hydrometeorological Prediction Center. Unified Surface Analysis Manual. Retrieved on 2006-10-22.
14. Saseendran S. A., Harenduprakash L., Rathore L. S. and Singh S. V. A GIS application for weather analysis and forecasting. Retrieved on 2007-05-05.
15. National Weather Service. Station Model Example.Retrieved on 2007-04-29. Archived October 25, 2007, at the Wayback Machine.
16. Dr Elizabeth R. Tuttle. Weather Maps. Retrieved on 2007-05-10.
17. American Meteorological Society. Selected DataStreme Atmosphere Weather Map Symbols. Retrieved on 2007-05-10.
18. CoCoRAHS. INTRODUCTION TO DRAWING ISOPLETHS. Retrieved on 2007-04-29. Archived April 28, 2007, at the Wayback Machine.
19. Glossary of meteorology. Synoptic scale. Retrieved on 2007-05-10.
20. Weather Doctor. Weather's Highs and Lows: Part 1 The High.
21. Agencia Estatal de Meteorología. Meteorología del aeropuerto de La Palma..
22. BBC Weather. Weather Basics - Low Pressure. Retrieved on 2007-05-05.
23. BBC Weather. High Pressure. Retrieved on 2007-05-05.
24. United Kingdom School System. Pressure, Wind and Weather Systems. Retrieved on 2007-05-05.
25. University of Illinois. Occluded Front. Retrieved on 2006-10-22.
26. National Weather Service Office, Norman, Oklahoma. Triple Point. Retrieved on 2006-10-22. Archived October 9, 2006, at the Wayback Machine.
27. "Trowal". World Meteorological Organisation. Eumetcal. Retrieved 2013-08-28.
28. University of Illinois. Stationary Front. Retrieved on 2006-10-22.
29. Glossary of Meteorology. Shear Line. Retrieved on 2006-10-22.
30. Fujita, T. T., 1986. Mesoscale classifications: their history and their application to forecasting. Mesoscale Meteorology and Forecasting. American Meteorological Society, Boston, p. 18–35.

31. Huaqing Cai. Dryline cross section. Retrieved on 2006-12-05.
32. "Lecture 3". Archived from the original on 27 September 2007.
33. Lewis D. Grasso. A Numerical Simulation of Dryline Sensitivity to Soil Moisture. Retrieved on 2007-05-10.
34. Glossary of Meteorology. Lee Trough. Retrieved on 2006-10-22.
35. University of Illinois. Dry Line: A Moisture Boundary.Retrieved on 2006-10-22.
36. Office of the Federal Coordinator for Meteorology.Chapter 2: Definitions. Retrieved on 2006-10-22.
37. Glossary of Meteorology. Sea Breeze. Retrieved on 2006-10-22.
38. http://www.climate.umn.edu/doc/journal/cold_december_2013.htm
39. David Roth (2006-12-14). "Unified Surface Analysis Manual" (PDF). Hydrometeorological Prediction Center. Retrieved 2012-01-09.
40. Paul M. Markowski; Yvette P. Richardson (2011-09-20). Mesoscale Meteorology in Midlatitudes. John Wiley and Sons. p. 120. ISBN 978-1-119-96667-8. Retrieved 2012-01-09.
41. Lee M. Grenci; Jon M. Nese (2001). A World of Weather: Fundamentals of Meteorology: A Text / Laboratory Manual (3 ed.). Kendall/Hunt Publishing Company. pp. 207–212. ISBN 978-0-7872-7716-1. OCLC 51160155.
42. Preview Text, Holton, James R., 2004: An Introduction to Dynamic Meteorology. Academic Press, 277 pp.
43. Weston, K. J. (1980). "An observational study of convective cloud streets". Tellus. 32 (5): 433. doi:10.1111/j.2153-3490.1980.tb00970.x.
44. Glossary of Meteorology (2009). Prefrontal squall line. Retrieved on 2008-12-24.
45. Schemm, S.; L. Nisi, A. Martinov; D. Leuenberg & O. Martius (2016). "On the link between cold fronts and hail in Switzerland". Atmospheric Science Letters. 17 (5): 315–325. doi:10.1002/asl.660.
46. K. A. Browning and Robert J. Gurney (1999). Global Energy and Water Cycles. Retrieved on 2008-12-26.
47. Meteorology Today p. 298, Ahrens, Donald C., 2007: Meteorology Today: an introduction to weather, climate, and the environment. Thomas Learning, Inc., 298pp.
48. "Overrunning". NWS Glossary. National Weather Service. Retrieved 2010-05-02.
49. Preview Text, Carlson, Toby N., 1991: Mid-Latitude Weather Systems. HarperCollins, 435 pp.
50. Nav Canada (January 2005). Aviation Weather Services Guide. p. 36.
51. St. Louis University (2003-08-04). "What is a TROWAL? via the Internet Wayback Machine". Archived from the original on 2006-09-16. Retrieved 2006-11-02.

CHAPTER 3

Weather Observation and Prediction of Weather

Chapter Contents

INTRODUCTION

Weather reconnaissance is the acquisition of weather data used for research and planning. Typically the term reconnaissance refers to observing weather from the air, as opposed to the ground (for example, ground based weather radar).

A WC-130 Weatherbird hurricane hunter

METHODS

Aircraft

Balloon

Weather balloon

Helicopter

Helicopters are not built to withstand the severe turbulence encountered in hurricane rainbands and eye walls. One reason is that a helicopter receives all of its lift from its rotating blades, and they are most likely to break off in hurricane conditions.

Fixed-Wing Aircraft

The Lockheed WC-130J aircraft is a venerable workhorse. It flies directly into the hurricane, typically penetrating the hurricane's eye several times per mission at altitudes between 500 feet (150 m) and 10,000 feet (3,000 m). The 53rd WRS Hurricane Hunters operate ten WC-130J aircraft for weather reconnaissance.

The WP-3D Orion aircraft flown by the NOAA Hurricane Hunters are heavily instrumented flying laboratories specifically modified to take atmospheric and radar measurements within tropical cyclones and winter storms.

The NOAA Gulfstream IV high altitude jet conducts hurricane surveillance flying upwards of 4,000 miles (6,400 km) each flight to document upper and lower level winds that affect the movement of tropical cyclones. The hurricane models (computer models predicting hurricane tracks and intensity) mainly utilize NOAA G-IV dropwindsonde data that is collected both day and night in storms affecting the United States.

Other aircraft have been used to investigate hurricanes, including an instrumented Lockheed U-2 that was flown in Hurricane Ginny during the 1963 Atlantic hurricane season.

Past aircraft used were the A-20 Havoc, 1944; B-24, 1944–1945; B-17, 1945–1947; B-25, 1946–1947; B-29, 1946–1947. WB-29, 1951–1956; WB-50, 1956–1963; WB-47, 1963–1969; WC-121N 1954-1973; WC-130A,B,E,H, 1965-2005.

Satellite

Watercraft

Watercraft deployed for use as weather ships have fallen out of favor due to their high operating cost. Unmanned weather buoys replaced weather ships when they became prohibitively expensive. Since the 1970s, their role has been largely superseded by weather buoys by design. Across the northern Atlantic, the number of weather ships dwindled over the years. The original nine ships in the region had fallen to eight by the 1970s. In 1974, the Coast Guard announced plans to terminate the United States stations, and, in 1977, the last United States weather ship was replaced by a newly developed weather buoy.

By 1983, data was still being collected by ships "M" ("Mike"), "R" ("Romeo"), "C" ("Charlie"), and L ("Lima"), Because of high operating costs and budget issues, weather ship "R" was recalled from the Bay of Biscay before the deployment of a weather buoy for the region. This recall was blamed for the minimal warning given in advance of the Great Storm of 1987.The last weather ship was *Polarfront*, known as weather station "M" at 66°N, 02°E, run by the Norwegian Meteorological Institute. *Polarfront' was put out of operation 1 January 2010. Despite the loss of designated weather ships, weather observations from ships continue from a fleet of voluntary merchant vessels in routine commercial operation, which have increased in number over the decades.*

APPLICATIONS

Military

Research

Spaceflight Planning

Images from satellites provide a resource for forecasting weather for NASA Space Shuttle launches and landings. Meteorologists analyze images to predict regions of cloud formation and dissipation. Special attention is paid to low clouds and convective cloud particularly cumulonimbus incus clouds. Satellite imagery is used to ascertain cloud-top temperatures to analyze the potential for lightning. Certain types of imagery are valued for their ability to view fog and low clouds at night. Satellite imagery in the long term can help enhance the shuttle flight landing procedure.

Prior to shuttle launches or landings, pilots fly aircraft that provide cloud, wind, turbulence, visibility, and precipitation information. Aircraft are flown along the future flight path of the shuttle and observations are noted. This complements radar and satellite data and only provides information that is useful for short-term (up to four hours before launch or landing) but not long-term forecasting. Aerial reconnaissance often provides a more accurate assessment of weather conditions than radar or satellite imagery.

Weather reconnaissance is also provided by weather balloons.

WEATHER FORECASTING

Weather forecasting is the application of science and technology to predict the state of the atmosphere for a given location. Human beings have attempted to predict the weather informally for millennia, and formally since the nineteenth century. Weather forecasts are made by collecting quantitative data about the current state of the atmosphere at a given place and using scientific understanding of atmospheric processes to project how the atmosphere will change.

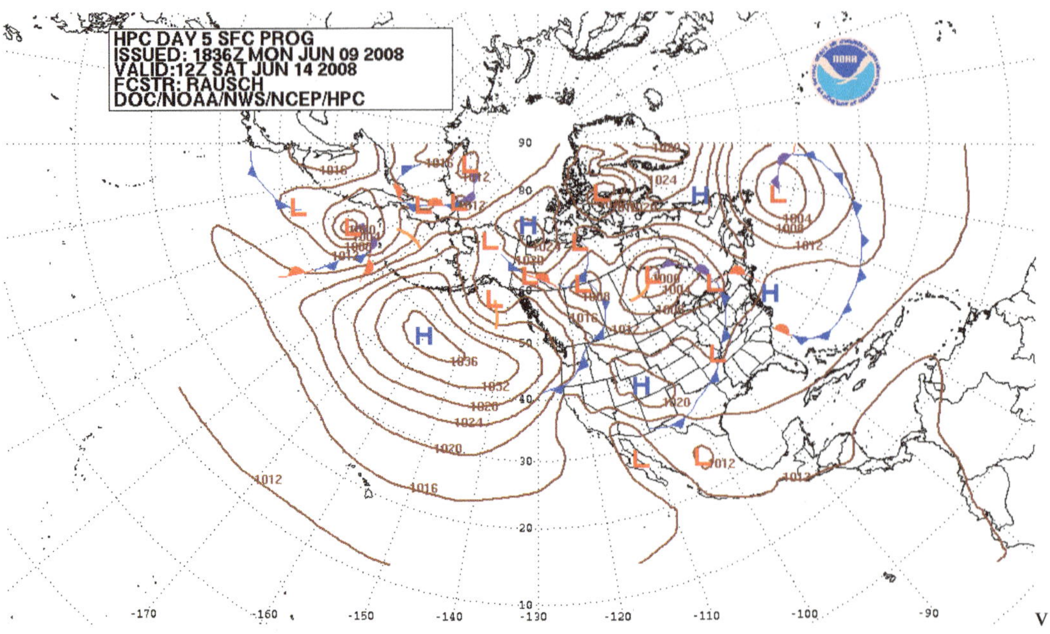

Forecast of surface pressures five days into the future for the north Pacific, North America, and north Atlantic Ocean

Once an all-human endeavor based mainly upon changes in barometric pressure, current weather conditions, and sky condition, weather forecasting now relies on computer-based models that take many atmospheric factors into account. Human input is still required to pick the best possible forecast model to base the forecast upon, which involves pattern recognition skills, teleconnections, knowledge of model performance, and knowledge of model biases. The inaccuracy of forecasting is due to the chaotic nature of the atmosphere, the massive computational power required to solve the equations that describe the atmosphere, the error involved in measuring the initial conditions, and an incomplete understanding of atmospheric processes. Hence, forecasts become less accurate as the difference between current time and the time for which the forecast is being made (the range of the forecast) increases. The use of ensembles and model consensus help narrow the error and pick the most likely outcome.

There are a variety of end uses to weather forecasts. Weather warnings are important forecasts because they are used to protect life and property. Forecasts based on temperature and precipitation are important to agriculture, and therefore to traders within commodity markets. Temperature forecasts are used by utility companies to estimate demand over coming days. On an everyday basis, people use weather forecasts to determine what to wear on a given day. Since outdoor activities are severely curtailed by heavy rain, snow and wind chill, forecasts can be used to plan activities around these events, and to plan ahead and survive them. In 2014, the US spent $5.1 billion on weather forecasting.

Ancient Forecasting

For millennia people have tried to forecast the weather. In 650 BC, the Babylonians predicted the weather from cloud patterns as well as astrology. In about 340 BC, Aristotle described weather patterns in *Meteorologica*. Later, Theophrastus compiled a book on weather forecasting, called the *Book of Signs*. Chinese weather prediction lore extends at least as far back as 300 BC, which was also around the same time ancient Indian astronomers developed weather-prediction methods. In New Testament times, Christ himself referred to deciphering and understanding local weather patterns, by saying, "When evening comes, you say, 'It will be fair weather, for the sky is red', and in the morning, 'Today it will be stormy, for the sky is red and overcast.' You know how to interpret the appearance of the sky, but you cannot interpret the signs of the times."

In 904 AD, Ibn Wahshiyya's *Nabatean Agriculture* discussed the weather forecasting of atmospheric changes and signs from the planetary astral alterations; signs of rain based on observation of the lunar phases; and weather forecasts based on the movement of winds.

Ancient weather forecasting methods usually relied on observed patterns of events, also termed pattern recognition. For example, it might be observed that if the sunset was particularly red, the following day often brought fair weather. This experience accumulated over the generations to produce weather lore. However, not all of these predictions prove reliable, and many of them have since been found not to stand up to rigorous statistical testing.

Modern Methods

The *Royal Charter* sank in an 1859 storm, stimulating the establishment of modern weather forecasting.

It was not until the invention of the electric telegraph in 1835 that the modern age of weather forecasting began. Before that, the fastest that distant weather reports could travel was around 100 miles per day (160 km/d), but was more typically 40–75 miles per day (60–120 km/day) (whether by land or by sea). By the late 1840s, the telegraph allowed reports of weather conditions from a wide area to be received almost instantaneously, allowing forecasts to be made from knowledge of weather conditions further upwind.

The two men credited with the birth of forecasting as a science were officer of the Royal Navy Francis Beaufort and his protégé Robert FitzRoy. Both were influential men in British naval and governmental circles, and though ridiculed in the press at the time, their work gained scientific credence, was accepted by the Royal Navy, and formed the basis for all of today›s weather forecasting knowledge.

Beaufort developed the Wind Force Scale and Weather Notation coding, which he was to use in his journals for the remainder of his life. He also promoted the development of reliable tide tables around British shores, and with his friend William Whewell, expanded weather record-keeping at 200 British Coast guard stations.

Robert FitzRoy was appointed in 1854 as chief of a new department within the Board of Trade to deal with the collection of weather data at sea as a service to mariners. This was the forerunner of the modern Meteorological Office. All ship captains were tasked with collating data on the weather and computing it, with the use of tested instruments that were loaned for this purpose.

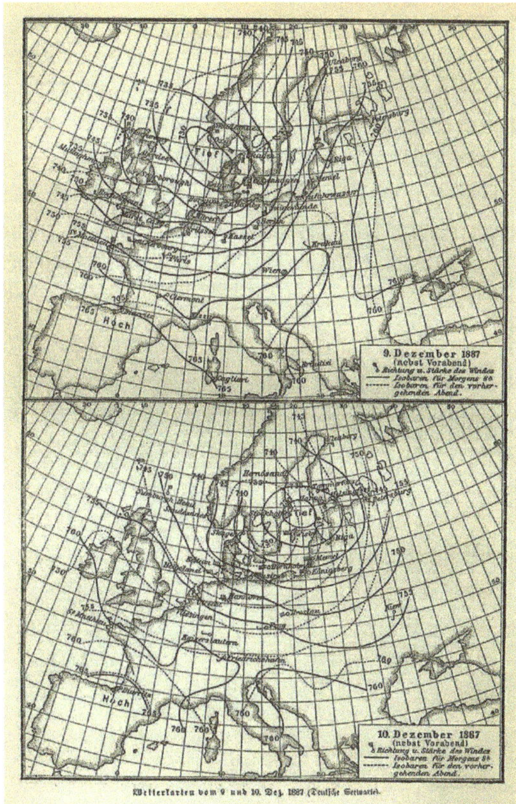

Weather map of Europe, 10 December 1887.

A storm in 1859 that caused the loss of the *Royal Charter* inspired FitzRoy to develop charts to allow predictions to be made, which he called *"forecasting the weather"*, thus coining the term "weather forecast". Fifteen land stations were established to use the new telegraph to transmit to him daily reports of weather at set times leading to the first gale warning service. His warning service for shipping was initiated in February 1861, with the use of telegraph communications. The first daily weather forecasts were published in *The Times* in 1861. In the following year a system was introduced of hoisting storm warning cones at the principal ports when a gale was expected. The *«Weather Book»* which FitzRoy published in 1863 was far in advance of the scientific opinion of the time.

As the electric telegraph network expanded, allowing for the more rapid dissemination of warnings, a national observational network was developed which could then be used to provide synoptic analyses. Instruments to continuously record variations in meteorological parameters using photography were supplied to the observing stations from Kew Observatory – these cameras had been invented by Francis Ronalds in 1845 and his barograph had earlier been used by FitzRoy.

To convey accurate information, it soon became necessary to have a standard vocabulary describing clouds; this was achieved by means of a series of classifications first achieved by Luke Howard in 1802, and standardized in the *International Cloud Atlas* of 1896.

Numerical Prediction

It was not until the 20th century that advances in the understanding of atmospheric physics led to the foundation of modern numerical weather prediction. In 1922, English scientist Lewis Fry Richardson published "Weather Prediction By Numerical Process", after finding notes and derivations he worked on as an ambulance driver in World War I. He described therein how small terms in the prognostic fluid dynamics equations governing atmospheric flow could be neglected, and a finite differencing scheme in time and space could be devised, to allow numerical prediction solutions to be found.

Richardson envisioned a large auditorium of thousands of people performing the calculations and passing them to others. However, the sheer number of calculations required was too large to be completed without the use of computers, and the size of the grid and time steps led to unrealistic results in deepening systems. It was later found, through numerical analysis, that this was due to numerical instability. The first computerised weather forecast was performed by a team led by the mathematician John von Neumann; von Neumann publishing the paper *Numerical Integration of the Barotropic Vorticity Equation* in 1950. Practical use of numerical weather prediction began in 1955, spurred by the development of programmable electronic computers.

Broadcasts

George Cowling (above) presented the first in-vision forecast on 11 January 1954 for the BBC.

The first ever daily weather forecasts were published in *The Times* on 1 August 1861, and the first weather maps were produced later in the same year. In 1911, the Met Office began issuing the first marine weather forecasts via radio transmission. These included gale and storm warnings for areas around Great Britain. In the United States, the first public radio forecasts were made in 1925 by Edward B. «E.B.» Rideout, on WEEI, the Edison Electric Illuminating station in Boston. Rideout came from the U.S. Weather Bureau, as did WBZ weather forecaster G. Harold Noyes in 1931.

The world›s first televised weather forecasts, including the use of weather maps, were experimentally broadcast by the BBC in 1936. This was brought into practice in 1949 after World War II. George Cowling gave the first weather forecast while being televised in front of the map in 1954. In America, experimental television forecasts were made by James C Fidler in Cincinnati in either 1940 or 1947 on the DuMont Television Network. In the late 1970s and early 80s, John Coleman, the first weatherman on ABC-TV›s Good Morning America, pioneered the use of on-screen weather satellite information and computer graphics for television forecasts. Coleman was a co-founder of The Weather Channel (TWC) in 1982. TWC is now a 24-hour cable network. Some weather channels have started broadcasting on live broadcasting programs such as YouTube and Periscope to reach more viewers. One Meteorologist, Jeff Matthews, transformed the medium from forecasting live during television broadcasts to forecasting live on Periscope under the name of «Weathertainment», becoming the first meteorologist to deliver daily forecasts on a life-streaming application...and thus, making weather forecasting, interactive, for the public.

How Models Create Forecasts

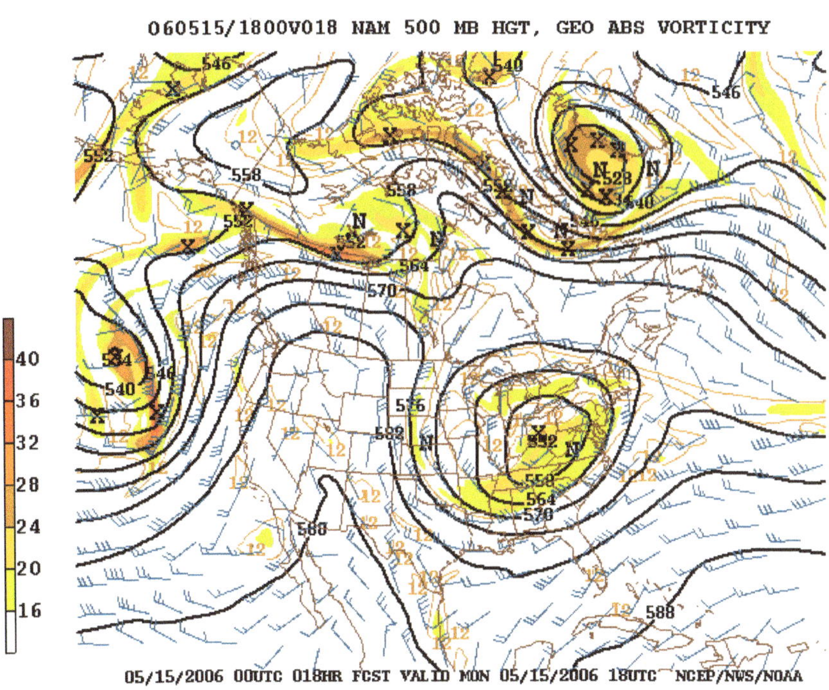

An example of 500 mbargeopotential height and absolute vorticity prediction from a numerical weather prediction model

The basic idea of numerical weather prediction is to sample the state of the fluid at a given time and use the equations of fluid dynamics and thermodynamics to estimate the state of the fluid at some time in the future. The main inputs from country-based weather services are surface observations from automated weather stations at ground level over land and from weather buoys at sea. The World Meteorological Organization acts to standardize the instrumentation, observing practices and timing of these observations worldwide. Stations either report hourly in METAR reports, or every six hours in SYNOP reports. Sites launch radiosondes, which rise through the depth of the troposphere and well into the stratosphere. Data from weather satellites are used in areas where traditional data sources are not available. Compared with similar data from radiosondes, the satellite data has the advantage of global coverage, however at a lower accuracy and resolution. Meteorological radar provide information on precipitation location and intensity, which can be used to estimate precipitation accumulations over time.Additionally, if a pulse Doppler weather radar is used then wind speed and direction can be determined.

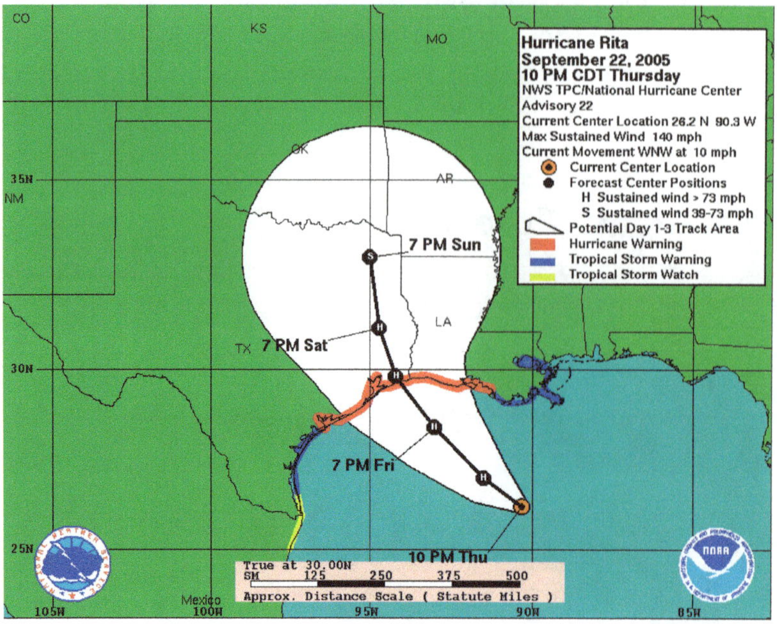

Modern weather predictions aid in timely evacuations and potentially save lives and prevent property damage

Commerce provides pilot reports along aircraft routes, and ship reports along shipping routes. Research flights using reconnaissance aircraft fly in and around weather systems of interest such as tropical cyclones. Reconnaissance aircraft are also flown over the open oceans during the cold season into systems which cause significant uncertainty in forecast guidance, or are expected to be of high impact 3–7 days into the future over the downstream continent.

Models are *initialized* using this observed data. The irregularly spaced observations are processed by data assimilation and objective analysis methods, which perform quality control and obtain values at locations usable by the model's mathematical algorithms (usually an evenly spaced grid). The data are then used in the model as the starting point for a forecast. Commonly, the set of equations used to predict the known as the physics and dynamics of the atmosphere are called primitive equations. These equations are initialized from the analysis data and rates of change are determined. The rates of change predict the state of the atmosphere a short time into the future. The equations are then applied to this new atmospheric state to find new rates of change, and these new rates of change predict the atmosphere at a yet further time into the future. This *time stepping* procedure is continually repeated until the solution reaches the desired forecast time. The length of the time step is related to the distance between the points on the computational grid.

The length of the time step chosen within the model is related to the distance between the points on the computational grid, and is chosen to maintain numerical stability. Time steps for global models are on the order of tens of minutes, while time steps for regional models are between one and four minutes. The global models are run at varying times into the future. The UKMET Unified Model is run six days into the future, the European Centre for Medium-Range Weather Forecasts model is run out to 10 days into the future, while the Global Forecast System model run by the Environmental Modeling Center is run 16 days into the future. The visual output produced by a model solution is known as a prognostic chart, or *prog*. The raw output is often modified before being presented as the forecast. This can be in the form of statistical techniques to remove known biases in the model, or of adjustment to take into account consensus among other numerical weather forecasts. MOS or model output statistics is a technique used to interpret numerical model output and produce site-specific guidance. This guidance is presented in coded numerical form, and can be obtained for nearly all National Weather Service reporting stations in the United States. As proposed by Edward Lorenz in 1963, long range forecasts, those made at a range of two weeks or more, are impossible to definitively predict the state of the atmosphere, owing to the chaotic nature of the fluid dynamics equations involved. In numerical models, extremely small errors in initial values double roughly every five days for variables such as temperature and wind velocity.

Essentially, a model is a computer program that produces meteorological information for future times at given locations and altitudes. Within any modern model is a set of equations, known as the primitive equations, used to predict the future state of the atmosphere. These equations—along with the ideal gas law—are used to evolve the density, pressure, and potential temperature scalar fields and the velocity vector field of the atmosphere through time. Additional transport equations for pollutants and other aerosols are included in some primitive-equation mesoscale models as well. The equations used are nonlinear partial differential equations which are impossible to solve exactly through analytical methods, with the exception of a few idealized cases. Therefore, numerical methods obtain approximate solutions. Different models use different solution methods: some global models use spectral methods for the horizontal dimensions and finite difference methods for the vertical dimension, while regional models and other global models usually use finite-difference methods in all three dimensions.

Techniques

Persistence

The simplest method of forecasting the weather, persistence, relies upon today's conditions to forecast the conditions tomorrow. This can be a valid way of forecasting the weather when it is in a steady state, such as during the summer season in the tropics. This method of forecasting strongly depends upon the presence of a stagnant weather pattern. It can be useful in both short range forecasts and long range forecasts.

Use of a Barometer

Measurements of barometric pressure and the pressure tendency (the change of pressure over time) have been used in forecasting since the late 19th century. The larger the change in pressure, especially if more than 3.5 hPa (2.6 mmHg), the larger the change in weather can be expected. If the pressure drop is rapid, a low pressure system is approaching, and there is a greater chance of rain. Rapid pressure rises are associated with improving weather conditions, such as clearing skies.

Looking at the Sky

Marestail shows moisture at high altitude, signalling the later arrival of wet weather.

Along with pressure tendency, the condition of the sky is one of the more important parameters used to forecast weather in mountainous areas. Thickening of cloud cover or the invasion of a higher cloud deck is indicative of rain in the near future. At night, high thin cirrostratus clouds can lead to halos around the moon, which indicates an approach of a warm front and its associated rain. Morning fog portends fair conditions, as rainy conditions are preceded by wind or clouds which prevent fog formation. The approach of a line of thunderstorms could indicate the approach of a cold front. Cloud-free skies are indicative of fair weather for the near future. A bar can indicate a coming tropical cyclone. The use of sky cover in weather prediction has led to various weather lore over the centuries.

Nowcasting

The forecasting of the weather within the next six hours is often referred to as nowcasting. In this time range it is possible to forecast smaller features such as individual showers and thunderstorms with reasonable accuracy, as well as other features too small to be resolved by a computer model. A human given the latest radar, satellite and observational data will be able to make a better analysis of the small scale features present and so will be able to make a more accurate forecast for the following few hours. However, there are now expert systems using those data and mesoscale numerical model to make better extrapolation, including evolution of those features in time.

Use of Forecast Models

In the past, the human forecaster was responsible for generating the entire weather forecast based upon available observations. Today, human input is generally confined to choosing a model based on various parameters, such as model biases and performance. Using a consensus of forecast models, as well as ensemble members of the various models, can help reduce forecast error. However, regardless how small the average error becomes with any individual system, large errors within any particular piece of guidance are still possible on any given model run. Humans are required to interpret the model data into weather forecasts that are understandable to the end user. Humans can use knowledge of local effects which may be too small in size to be resolved by the model to add information to the forecast. While increasing accuracy of forecast models implies that humans may no longer be needed in the forecast process at some point in the future, there is currently still a need for human intervention.

Analog Technique

The analog technique is a complex way of making a forecast, requiring the forecaster to remember a previous weather event which is expected to be mimicked by an upcoming event. What makes it a difficult technique to use is that there is rarely a perfect analog for an event in the future. Some call this type of forecasting pattern recognition. It remains a useful method of observing rainfall over data voids such as oceans, as well as the forecasting of precipitation amounts and distribution in the future. A similar technique is used in medium range forecasting, which is known as teleconnections, when systems in other locations are used to help pin down the location of another system within the surrounding regime. An example of teleconnections are by using El Niño-Southern Oscillation (ENSO) related phenomena.

Quality Assessment

Mathematical models are often developed to map the forecast skill of a weather prediction methodology.

Communicating Forecasts to the Public

An example of a two-day weather forecast in the visual style that an American newspaper might use. Temperatures are given in Fahrenheit.

Most end users of forecasts are members of the general public. Thunderstorms can create strong winds and dangerous lightning strikes that can lead to deaths, power outages, and widespread hail damage. Heavy snow or rain can bring transportation and commerce to a stand-still, as well as cause flooding in low-lying areas. Excessive heat or cold waves can sicken or kill those with inadequate utilities, and droughts can impact water usage and destroy vegetation.

Several countries employ government agencies to provide forecasts and watches/warnings/advisories to the public in order to protect life and property and maintain commercial interests. Knowledge of what the end user needs from a weather forecast must be taken into account to present the information in a useful and understandable way. Examples include the National Oceanic and Atmospheric Administration's National Weather Service (NWS) and Environment Canada's Meteorological Service (MSC). Traditionally, newspaper, television, and radio have been the primary outlets for presenting weather forecast information to the public. Increasingly, the internet is being used due to the vast amount of specific information that can be found. In all cases, these outlets update their forecasts on a regular basis.

Severe Weather Alerts and Advisories

A major part of modern weather forecasting is the severe weather alerts and advisories which the national weather services issue in the case that severe or hazardous weather is expected. This is done to protect life and property. Some of the most commonly known of severe weather advisories are the severe thunderstorm and tornado warning, as well as the severe thunderstorm and tornado watch. Other forms of these advisories include winter weather, high wind, flood, tropical cyclone, and fog. Severe weather advisories and alerts are broadcast through the media, including radio, using emergency systems as the Emergency Alert System which break into regular programming.

Low Temperature Forecast

The low temperature forecast for the current day is calculated using the lowest temperature found between 7pm that evening through 7am the following morning. So, in short, today's forecasted low is most likely tomorrow's low temperature.

Specialist Forecasting

There are a number of sectors with their own specific needs for weather forecasts and specialist services are provided to these users.

Air Traffic

Ash cloud from the 2008 eruption of Chaitén volcano stretching across Patagonia from the Pacific to the Atlantic Ocean

Because the aviation industry is especially sensitive to the weather, accurate weather forecasting is essential. Fog or exceptionally low ceilings can prevent many aircraft from landing and taking off. Turbulence and icing are also significant in-flight hazards. Thunderstorms are a problem for all aircraft because of severe turbulence due to their updrafts and outflow boundaries, icing due to the heavy precipitation, as well as large hail, strong winds, and lightning, all of which can cause severe damage to an aircraft in flight. Volcanic ash is also a significant problem for aviation, as aircraft can lose engine power within ash clouds. On a day-to-day basis airliners are routed to take advantage of the jet stream tailwind to improve fuel efficiency. Aircrews are briefed prior to takeoff on the conditions to expect en route and at their destination. Additionally, airports often change which runway is being used to take advantage of a headwind. This reduces the distance required for takeoff, and eliminates potential crosswinds.

Marine

Commercial and recreational use of waterways can be limited significantly by wind direction and speed, wave periodicity and heights, tides, and precipitation. These factors can each influence the safety of marine transit. Consequently, a variety of codes have been established to efficiently transmit detailed marine weather forecasts to vessel pilots via radio, for example the MAFOR (marine forecast). Typical weather forecasts can be received at sea through the use of RTTY, Navtex and Radiofax.

Agriculture

Farmers rely on weather forecasts to decide what work to do on any particular day. For example, drying hay is only feasible in dry weather. Prolonged periods of dryness can ruin cotton, wheat, and corn crops. While corn crops can be ruined by drought, their dried remains can be used as a cattle feed substitute in the form of silage. Frosts and freezes play havoc with crops both during the spring and fall. For example, peach trees in full bloom can have their potential peach crop decimated by a spring freeze. Orange groves can suffer significant damage during frosts and freezes, regardless of their timing.

Forestry

Weather forecasting of wind, precipitations and humidity is essential for preventing and controlling wildfires. Different indices, like the *Forest fire weather index* and the *Haines Index*, have been developed to predict the areas more at risk to experience fire from natural or human causes. Conditions for the development of harmful insects can be predicted by forecasting the evolution of weather, too.

Utility Companies

Electricity and gas companies rely on weather forecasts to anticipate demand which can be strongly affected by the weather. They use the quantity termed the degree day to determine how strong of a use there will be for heating (heating degree day) or cooling (cooling degree day). These quantities are based on a daily average temperature of 65 °F (18 °C). Cooler temperatures force heating degree days (one per degree Fahrenheit), while warmer temperatures force cooling degree days. In winter, severe cold weather can cause a surge in demand as people turn up their heating. Similarly, in summer a surge in demand can be linked with the increased use of air conditioning systems in hot weather. By anticipating a surge in demand, utility companies can purchase additional supplies of power or natural gas before the price increases, or in some circumstances, supplies are restricted through the use of brownouts and blackouts.

An air handling unit is used for the heating and cooling of air in a central location (click on image for legend).

Other Commercial Companies

Increasingly, private companies pay for weather forecasts tailored to their needs so that they can increase their profits or avoid large losses.For example, supermarket chains may change the stocks on their shelves in anticipation of different consumer spending habits in different weather conditions. Weather forecasts can be used to invest in the commodity market, such as futures in oranges, corn, soybeans, and oil.

Military Applications

United Kingdom Armed Forces

Royal Navy

The UK Royal Navy, working with the UK Met Office, has its own specialist branch of weather observers and forecasters, as part of the Hydrographic and Meteorological (HM) specialisation, who monitor and forecast operational conditions across the globe, to provide accurate and timely weather and oceanographic information to submarines, ships and Fleet Air Arm aircraft.

Royal Air Force

A mobile unit in the RAF, working with the UK Met Office, forecasts the weather for regions in which British, allied servicemen and women are deployed. A group based at Camp Bastion provides forecasts for the British armed forces in Afghanistan.

United States Armed Forces

US Navy

Similar to the private sector, military weather forecasters present weather conditions to the war fighter community. Military weather forecasters provide pre-flight and in-flight weather briefs to pilots and provide real time resource protection services for military installations. Naval forecasters cover the waters and ship weather forecasts. The United States Navy provides a special service to both themselves and the rest of the federal government by issuing forecasts for tropical cyclones across the Pacific and Indian Oceans through their Joint Typhoon Warning Center.

US Air Force

Within the United States, Air Force Weather provides weather forecasting for the Air Force and the Army. Air Force forecasters cover air operations in both wartime and peacetime operations and provide Army support; United States Coast Guard marine science technicians provide ship forecasts for ice breakers and other various operations within their realm; and Marine forecasters provide support for ground- and air-based United States Marine Corps operations. All four military branches take their initial enlisted meteorology technical training at Keesler Air Force Base. Military and civilian forecasters actively cooperate in analyzing, creating and critiquing weather forecast products.

REFERENCES

1. "FAQ". Retrieved 6 February 2011.
2. J. F. Robin McIlveen (1998). Fundamentals of weather and climate. Psychology Press. p. 31. ISBN 978-0-7487-4079-6. Retrieved 2011-01-18.
3. National Research Council (U.S.). Ocean Science Committee, National Research Council (U.S.). Study Panel on Ocean Atmosphere Interaction (1974). The role of the ocean in predicting climate: a report of workshops conducted by Study Panel on Ocean Atmosphere Interaction under the auspices of the Ocean Science Committee of the Ocean Affairs Board, Commission on Natural Resources, National Research Council. National Academies. p. 40. Retrieved 2011-01-18.
4. Robertson P. Dinsmore (December 1996). "Alpha, Bravo, Charlie... Ocean Weather Ships 1940-1980". Woods Hole Oceanographic Institution Marine Operations. Retrieved 2011-01-31.
5. Pan-European Infrastructure for Ocean & Marine Data Management (2010-09-11). "North Atlantic Ocean Weather Ship (OWS) Surface Meteorological Data (1945-1983)". British Oceanographic Data Centre. Retrieved 2011-01-31.
6. "Romeo Would Have Spied the Storm". New Scientist. IPC Magazines. 116 (1583): 22. 1987-10-22. Retrieved 2011-01-18.
7. Frank C. Brody; Richard A. Lafosse; Dan G. Bellue; Timothy D. Oram (1997).

"Operations of the National Weather Service Spaceflight Meteorology Group". American Meteorological Society. Retrieved 6 February 2011.

8. Dirmeyer, Paul A.; Schlosser, C. Adam; Brubaker, Kaye L. (February 1, 2009). "Precipitation, Recycling, and Land Memory: An Integrated Analysis". JOURNAL OF HYDROMETEOROLOGY. 10: 278–288. doi:10.1175/2008JHM1016.1. Retrieved December 30, 2016.
9. Fostering Innovation, Creating Jobs, Driving Better Decisions: The Value of Government Data. Economics and Statistics Administration Office of the Chief Economist. July 2014. p. 15.
10. Meteorology by Lisa Alter
11. Weather: Forecasting from the Beginning
12. University of California Museum of Paleontology. Aristotle (384-322 B.C.E.). Retrieved on 2008-01-12.
13. David Pingree. "THE INDIAN AND PSEUDO-INDIAN PASSAGES IN GREEK AND LATIN ASTRONOMICAL AND ASTROLOGICAL TEXTS" (PDF). pp. 141–195 143–4. Retrieved 2010-03-01.
14. "Bible Gateway passage: Matthew 16:2-3 - English Standard Version". Bible Gateway. Retrieved 2016-12-01.
15. Fahd, Toufic. : 842. Missing or empty |title= (help), in Rashed, Roshdi; Morelon, Régis (1996). Encyclopedia of the History of Arabic Science. 3. Routledge. pp. 813–852. ISBN 0-415-12410-7.
16. Jerry Wilson. Skywatch Signs of the Weather. Retrieved on 2007-04-15.
17. "Joseph Henry: Inventor of the Telegraph? Smithsonian Institution". Archived from the original on June 26, 2006. Retrieved 2006-06-29.
18. Ausman, Megaera. "USPS Historian". About the United States Postal Service. USPS. Retrieved 28 April 2013.
19. Mail, Royal. "(UK)". British Postal Museum. Postal Heritage Trust. Retrieved 28 April 2013.
20. Encyclopædia Britannica. Telegraph. Retrieved on 2007-05-05.
21. Eric D. Craft. "An Economic History of Weather Forecasting". Archived from the original on 3 May 2007. Retrieved 15 April 2007.
22. "The birth of the weather forecast". BBC. 30 April 2015. Retrieved 30 April 2015.
23. Mellersh, H. E. L. (1968). FitzRoy of the Beagle. Hart-Davis. ISBN 0-246-97452-4
24. Kington, John (1997). Mike Hulme and Elaine Barrow, ed. Climates of the British Isles: Present, Past and Future. Routledge. p. 147.
25. Ronalds, B.F. (2016). Sir Francis Ronalds: Father of the Electric Telegraph. London: Imperial College Press. ISBN 978-1-78326-917-4.

26. Ronalds, B.F. (June 2016). "Sir Francis Ronalds and the Early Years of the Kew Observatory". Weather. doi:10.1002/wea.2739.
27. Richardson, Lewis Fry, Weather Prediction by Numerical Process (Cambridge, England: Cambridge University Press, 1922). Available on-line at: Internet Archive.org.
28. Lynch, Peter (2006). The Emergence of Numerical Weather Prediction. Cambridge U.P.
29. Charney, Fjörtoft and von Neumann, 1950, Numerical Integration of the Barotropic Vorticity Equation Tellus, 2, 237-254
30. Paul N. Edwards. Atmospheric General Circulation Modeling. Retrieved on 2007-02-16.
31. Helen Czerski. "Orbit: Earth's Extraordinary Journey: 150 years since the first UK weather "forecast"". BBC. Retrieved November 5, 2013.
32. Met Office (2012). "National Meteorological Library and Fact Sheet 8 -- The Shipping Forecast" (PDF). 1. pp. 3–5. Retrieved 2013-04-10.
33. "meteorology Facts, information, pictures | Encyclopedia.com articles about meteorology". Encyclopedia.com. Retrieved 2014-02-21.
34. BBC - Weather - A history of TV weather forecasts BBC Weather
35. Hunt, Roger, "The end of weather forecasting at Met Office London", Weather, June 2007, v.62, no.6, pp.143-146
36. "Answers: Understanding weather forecasts". USA Today. 2006-02-08.
37. CJR Rewind: Hot Air, Columbia Journalism Review, reprint, first published in the January/February 2010 issue.
38. National Climatic Data Center. Key to METAR Surface Weather Observations. Retrieved on 2008-03-09.
39. UNISYS. SYNOP Data Format (FM-12): Surface Synoptic Observations. Retrieved on 2008-05-25.
40. Gaffen, Dian J. (2007-06-07). Radiosonde Observations and Their Use in SPARC-Related Investigations. Retrieved on 2008-05-25.
41. NASA. Interactive Global Composite Weather Satellite Images. Retrieved on 2008-05-25.
42. NOAA. Goes Eastern US Sector Infrared Image.Retrieved on 2008-05-25.
43. Met Office. Satellite applications. Retrieved on 2008-05-25.
44. Tony Reale. ATOVS Sounding Products (ITSVC-12).Retrieved on 2008-05-25.
45. Andrew Treloar and Peter Brookhouse. The use of accumulated rainfall maps from weather radar systems to assist wildfire detection reconnaissance. Retrieved on 2008-05-25.
46. University of Washington. An improving forecast.Retrieved on 2007-04-15
47. Ballish, Bradley A. and V. Krishna Kumar (2008-05-23). Investigation of

Systematic Differences in Aircraft and Radiosonde Temperatures with Implications for NWP and Climate Studies. Retrieved on 2008-05-25.

48. 403rd Wing (2011). "The Hurricane Hunters". 53rd Weather Reconnaissance Squadron. Retrieved 2006-03-30.
49. Lee, Christopher. "Drone, Sensors May Open Path Into Eye of Storm". The Washington Post. Retrieved 2008-02-22.
50. National Oceanic and Atmospheric Administration (2010-11-12). "NOAA Dispatches High-Tech Research Plane to Improve Winter Storm Forecasts". Retrieved 2010-12-22.
51. University Corporation for Atmospheric Research (2007-08-14). The WRF Variational Data Assimilation System (WRF-Var). Retrieved on 2008-05-25.
52. Pielke, Roger A. (2002). Mesoscale Meteorological Modeling. Academic Press. pp. 285–287. ISBN 0-12-554766-8.
53. Sunderam, V. S.; van Albada, G. Dick; Peter, M. A.; Sloot, J. J. Dongarra (2005). Computational Science – ICCS 2005: 5th International Conference, Atlanta, GA, USA, May 22–25, 2005, Proceedings, Part 1. Springer. p. 132. ISBN 978-3-540-26032-5. Retrieved 2011-01-02.
54. Zwieflhofer, Walter; Kreitz, Norbert; European Centre for Medium Range Weather Forecasts (2001). Developments in teracomputing: proceedings of the ninth ECMWF Workshop on the Use of High Performance Computing in Meteorology. World Scientific. p. 276. ISBN 978-981-02-4761-4. Retrieved 2011-01-02.
55. Chan, Johnny C. L. & Jeffrey D. Kepert (2010). Global Perspectives on Tropical Cyclones: From Science to Mitigation. World Scientific. pp. 295–296. ISBN 978-981-4293-47-1. Retrieved 2011-02-24.
56. Holton, James R. (2004). An introduction to dynamic meteorology, Volume 1. Academic Press. p. 480. ISBN 978-0-12-354015-7. Retrieved 2011-02-24.
57. Brown, Molly E. (2008). Famine early warning systems and remote sensing data. Springer. p. 121. ISBN 978-3-540-75367-4. Retrieved 2011-02-24.
58. Ahrens, C. Donald (2008). Essentials of meteorology: an invitation to the atmosphere. Cengage Learning. p. 244. ISBN 978-0-495-11558-8. Retrieved 2011-02-11.
59. Daniel Andersson. Improved accuracy of surrogate models using output postprocessing. Retrieved on 2008-05-25.
60. Cox, John D. (2002). Storm Watchers. John Wiley & Sons, Inc. pp. 222–224. ISBN 0-471-38108-X.
61. Pielke, Roger A. (2002). Mesoscale Meteorological Modeling. Academic Press. pp. 48–49. ISBN 0-12-554766-8.
62. Pielke, Roger A. (2002). Mesoscale Meteorological Modeling. Academic Press. pp. 18–19. ISBN 0-12-554766-8.
63. Strikwerda, John C. (2004). Finite difference schemes and partial differential equations. SIAM. pp. 165–170. ISBN 978-0-89871-567-5. Retrieved 2010-12-31.

64. Pielke, Roger A. (2002). Mesoscale Meteorological Modeling. Academic Press. p. 65. ISBN 0-12-554766-8.
65. University of Illinois at Urbana-Champaign. Persistence Forecasting: Today equals Tomorrow. Retrieved on 2007-02-16.
66. USA Today. Understanding air pressure. Retrieved on 2008-05-25.
67. Weather Doctor. Applying The Barometer To Weather Watching. Retrieved on 2008-05-25.
68. Dennis Eskow (March 1983). "Make Your Own Weather Forecasts". Popular Mechanics. 159 (3): 148. Retrieved 2011-04-02.
69. Mark Moore (2009-03-25). Field Forecasting - A Short Summary. Retrieved on 2012-02-15.
70. Jerry Wilson. Skywatch: Signs of the Weather. Retrieved on 2008-05-25.
71. Glossary of Meteorology. Retrieved on 2015-05-26.
72. E-notes.com. Weather and Climate | What Is Nowcasting?Retrieved on 2011-09-08.
73. NASA. Weather Forecasting Through the Ages. Retrieved on 2008-05-25.
74. Klaus Weickmann, Jeff Whitaker, Andres Roubicek and Catherine Smith (2001-12-01). The Use of Ensemble Forecasts to Produce Improved Medium Range (3-15 days) Weather Forecasts. Climate Diagnostics Center. Retrieved on 2007-02-16.
75. Todd Kimberlain. Tropical cyclone motion and intensity talk (June 2007). Retrieved on 2007-07-21.
76. Richard J. Pasch, Mike Fiorino, and Chris Landsea. TPC/NHC'S Review of the NCEP Production Suite for 2006.Retrieved on 2008-05-05.
77. Roebber P. J. and Bosart L. F. The complex relationship between forecast skill and forecast value : A real-world analysis. Retrieved on 2008-05-25.
78. Other Forecasting Methods: climatology, analogue and numerical weather prediction. Retrieved on 2006-02-16.
79. Kenneth C. Allen. Pattern Recognition Techniques Applied to the NASA-ACTS Order-Wire Problem. Retrieved on 2007-02-16.
80. Weather Associates, Inc. The Role of Teleconnections & Ensemble Forecasting in Extended- to Medium-Range Forecasting. Retrieved on 2007-02-16.
81. Thinkquest.org. Teleconnections: Linking El Niño with Other Places. Retrieved on 2007-02-16.
82. University of Illinois at Urbana-Champaign. Lightning.Retrieved on 2007-02-16.
83. Associated Press. Upstate N.Y. residents dig out from heavy snow. Retrieved on 2008-05-25.
84. National Flood Insurance Program. Flood Risk Scenarios: Flash Flood. Retrieved on 2008-05-25.
85. National Weather Service. About NOAA's National Weather Service. Retrieved on 2007-02-16.

86. Environment Canada. Main website. Retrieved on 2007-02-16.
87. Canadian Heritage. Primary Sources of Local Information.Retrieved on 2008-05-26.
88. National Weather Service. National Weather Service Mission Statement. Retrieved on 2008-05-25.
89. Environment Canada. Canadian Weather Alerts.Retrieved on 2008-05-26.
90. Federal Communications Commission. Emergency Alert System. Retrieved on 2008-05-26.
91. Weather Channel - Calculation of Low Temperature Forecast
92. Government Printing Office. Title 14: Aeronautics and Space. Retrieved on 2008-05-26.
93. Aircraft Owners and Pilots Association. Aircraft Icing.Retrieved on 2008-05-26.
94. National Weather Service Forecast Office Dodge City, Kansas. Aviation Hazards They Didn't Tell You About.Retrieved on 2008-05-26.
95. Bureau of Meteorology. Aviation Hazards: Thunderstorms and Deep Convection. Retrieved on 2008-05-26.
96. Volcanic Ash Aviation Hazard. Retrieved on 2008-05-26.
97. Ned Rozell. Amazing flying machines allow time travel.Retrieved on 2008-05-08.
98. National Weather Service. A Pilot's Guide to Aviation Weather Services. Retrieved on 2008-05-26.
99. Eric C. King. Takeoff Tools Crosswind Calculator Instructions. Retrieved on 2008-05-26.
100. Great Lakes and Seaway Shipping. MAFOR Weather Code. Retrieved on 2008-05-27.
101. Blair Fannin. Dry weather conditions continue for Texas.Retrieved on 2008-05-26.
102. Dr. Terry Mader. Drought Corn Silage. Retrieved on 2008-05-26.
103. Kathryn C. Taylor. Peach Orchard Establishment and Young Tree Care. Retrieved on 2008-05-26.
104. Associated Press. After Freeze, Counting Losses to Orange Crop. Retrieved on 2008-05-26.
105. Climate Prediction Center. Degree Day Explanation.Retrieved on 2008-05-25.
106. The New York Times. Futures/Options; Cold Weather Brings Surge in Prices of Heating Fuels. Retrieved on 2008-05-25.
107. BBC. Heatwave causes electricity surge. Retrieved on 2008-05-25.
108. Toronto Catholic Schools. The Seven Key Messages of the Energy Drill Program. Retrieved on 2008-05-25.
109. CSIRO. Providing specialized weather forecasts.Retrieved on 2008-05-25.
110. Stephen Jewson and Rodrigo Caballero. The Use of Weather Forecasts in the Pricing of Weather Derivatives.Retrieved on 2008-05-25.

111. Met Office. Weather forecasting for military operations.Retrieved on 2012-10-23.
112. Joint Typhoon Warning Center. Joint Typhoon Warning Center Mission Statement. Retrieved on 2008-05-27.
113. United States Air Force.Air Force Weather Agency.Retrieved on 2008-05-26.
114. United States Military. US Coast Guard Jobs - Enlisted Occupations. Retrieved on 2008-05-26.
115. Rod Powers. United States Marine Corps Enlisted Job Descriptions and Qualification Factors: Field 68 - Meteorology and Oceanography (METOC). Retrieved on 2008-05-26.
116. Keesler Air Force Base. Military officers usually received their education from a civilian institution. Keesler News: March 9, 2006. United States Air Force Retrieved on 2008-05-26.

CHAPTER 4

Weather Map

Chapter Contents

The earliest weather charts were made by collecting synchronous weather reports by mail. However, it was not until 1816 that German physicist Heinrich Wilhelm Brandes created the first weather maps, which were hand drawn and reconstructed from data collected in 1783. The first telegraphic collection of synoptic meteorological reports and their mapping for forecasting was accomplished by Urbain-J.-J. Le Verrier during the mid-1800s.

INTRODUCTION

A weather map displays various Meteorological features across a particular area at a particular point in time and has various symbols which all have specific meanings. Such maps have been in use since the mid-19th century and are used for research and weather forecasting purposes. Maps using isotherms show temperature gradients, which can help locate weather fronts. Isotach maps, analyzing lines of equal wind speed, on a constant pressure surface of 300 mb or 250 mb show where the jet stream is located. Use of constant pressure charts at the 700 and 500 hPa level can indicate tropical cyclone motion. Two-dimensional streamlines based on wind speeds at various levels show areas of convergence and divergence in the wind field, which are helpful in determining the location of features within the wind pattern. A popular type of surface weather map is the surface weather analysis, which plots isobars to depict areas of high pressure and low pressure. Cloud codes are translated into symbols and plotted on these maps along with other meteorological data that are included in synoptic reports sent by professionally trained observers.

Sir Francis Galton, the inventor of the weather map.

The use of weather charts in a modern sense began in the middle portion of the 19th century in order to devise a theory on storm systems.During the Crimean War a storm devastated the French fleet at Balaklava, and the French scientist Urbain Le Verrier was able to show that if a chronological map of the storm had been issued, the path it would take could have been predicted and avoided by the fleet.

In England, the scientist Francis Galton heard of this work, as well as the pioneering weather forecasts of Robert Fitzroy. After gathering information from weather stations across the country for the month of October 1861, he plotted the data on a map using his own system of symbols, thereby creating the world›s first weather map. He used his map to prove that air circulated clockwise around areas of high pressure; he coined the term ‹anticyclone› to describe the phenomenon. He was also instrumental in publishing the first weather map in a newspaper, for which he modified the pantograph (an instrument for copying drawings) to inscribe the map onto printing blocks. *The Times* began printing weather maps using these methods with data from the Meteorological Office.

The introduction of country-wide weather maps required the existence of national telegraph networks so that data from across the country could be gathered in real time and remain relevant for subsequent analysis. The first such use of the telegraph for gathering data on the weather was the *Manchester Examiner* newspaper in 1847:

...led us to inquire if the electric telegraph was yet extended far enough from Manchester to obtain information from the eastern counties...inquiries were made at the following places; and answers were returned, which we append...

It was also important for time to be standardized across time zones so that the information on the map should accurately represent the weather at a given time. A standardized time system was first used to coordinate the British railway network in 1847, with the inauguration of Greenwich Mean Time.

In the USA, The Smithsonian Institution developed its network of observers over much of the central and eastern United States between the 1840s and 1860s once Joseph Henry took the helm. The U.S. Army Signal Corps inherited this network between 1870 and 1874 by an act of Congress, and expanded it to the west coast soon afterwards. At first, not all the data on the map was used due to a lack of time standardization. The United States fully adopted time zones in 1905, when Detroit finally established standard time.

20th Century

Light tables were important to the construction of surface weather analyses into the 1990s

The use of frontal zones on weather maps began in the 1910s in Norway. Polar front theory is attributed to Jacob Bjerknes, derived from a coastal network of observation sites in Norway during World War I. This theory proposed that the main inflow into a cyclone was concentrated along two lines of convergence, one ahead of the low and another trailing behind the low. The convergence line ahead of the low became known as either the steering line or the warm front. The trailing convergence zone was referred to as the squall line or cold front. Areas of clouds and rainfall appeared to be focused along these convergence zones. The concept of frontal zones led to the concept of air masses. The nature of the three-dimensional structure of the cyclone would wait for the development of the upper air network during the 1940s. Since the leading edge of air mass changes bore resemblance to the military fronts of World War I, the term "front" came into use to represent these lines. The United States began to formally analyze fronts on surface analyses in late 1942, when the WBAN Analysis Center opened in downtown Washington, D.C.

In addition to surface weather maps, weather agencies began to generate constant pressure charts. In 1948, the United States began the Daily Weather Map series, which at first analyzed the 700 hPa level, which is around 3,000 metres (9,800 ft) above sea level. By May 14, 1954, the 500 hPa surface was being analyzed, which is about 5,520 metres (18,110 ft) above sea level. The effort to automate map plotting began in the United States in 1969, with the process complete in the 1970s. Hong Kong completed their process of automated surface plotting by 1987.

By 1999, computer systems and software had finally become sophisticated enough to allow for the ability to underlay on the same workstation satellite imagery, radar imagery, and model-derived fields such as atmospheric thickness and frontogenesis in combination with surface observations to make for the best possible surface analysis. In the United States, this development was achieved when Intergraph workstations were replaced by n-AWIPS workstations. By 2001, the various surface analyses done within the National Weather Service were combined into the Unified Surface Analysis, which is issued every six hours and combines the analyses of four different centers. Recent advances in both the fields of meteorology and geographic information systems have made it possible to devise finely tailored products that take us from the traditional weather map into an entirely new realm. Weather information can quickly be matched to relevant geographical detail. For instance, icing conditions can be mapped onto the road network. This will likely continue to lead to changes in the way surface analyses are created and displayed over the next several years.

STATION MODEL

A station model is a symbolic illustration showing the weather occurring at a given reporting station. Meteorologists created the station model to fit a number of weather elements into a small space on weather maps. This allows map users to analyze patterns in air pressure, temperature, wind, cloud cover, and precipitation.

Station model plots use an internationally accepted coding a website convention that has changed little since August 1, 1941. Elements in the plot show the key weather elements, including temperature, dew point, wind, cloud cover, air pressure, pressure tendency, and precipitation.

Measurement Location and Units

Weather maps primarily use the station model to show surface weather conditions, but the model can also show the weather aloft as reported by a weather balloon's radiosonde or a pilot's report.

Station model plots use a mixture of metric and Imperial units depending on the map's location and what is being shown. Surface maps in the United States primarily use Imperial units, such as inches, degrees Fahrenheit, and knots. Most of the world, however, uses metric measurements for everything but wind speed, which is shown in knots.

Plotted Winds

The station model uses a wind barb to show both wind direction and speed. The wind barb shows the speed using "flags" on the end.

- Each half of a flag depicts five knots
- Each full flag depicts 10 knots
- Each pennant (filled triangle) depicts 50 knots

Winds are depicted as blowing from the direction the flags are facing. Therefore, a northeast wind will be depicted with a line extending from the cloud circle to the northeast, with flags indicating wind speed on the northeast end of this line. Once plotted on a map, an analysis of isotachs (lines of equal wind speeds) can be accomplished. Isotachs are particularly useful in diagnosing the location of the jet stream on upper level constant pressure charts, usually at or above the 300 hPa level.

The flags and pennants point to the low pressure, so it is possible to determine at which hemisphere the station is standing. The barbs in the figure at the right are located at the Northern Hemisphere, because the wind is circling counter clock-wise around a low-pressure area at the Northern Hemisphere (the wind is blowing in the opposite direction at the Southern Hemisphere, see also Buys Ballot›s law).

More than a century ago, winds were initially plotted as arrows facing downwind, with feathers on both sides of the staff to indicate wind speed. In the United States, the change to the modern convention of flags shown on one side of the staff to indicate wind speed took effect on August 1, 1941.

Cloud Cover

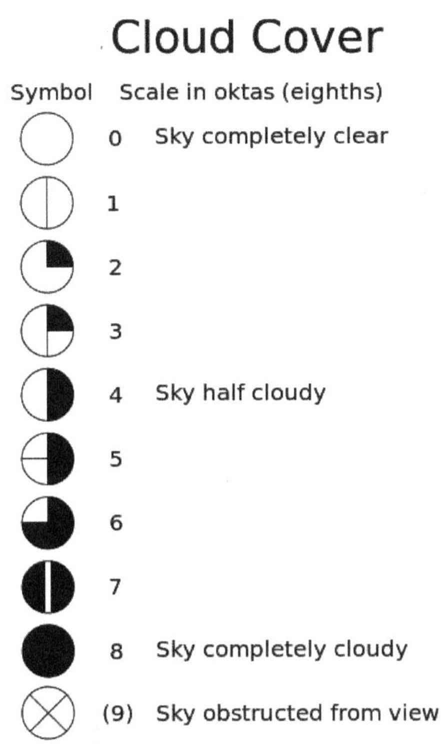

Along with wind direction, cloud cover is one of the oldest atmospheric conditions to be coded on a station model. The circle in the middle of the station model represents cloud cover. In the United Kingdom, when the observation is taken from an automated weather observation site, the shape is a triangle. If the shape is completely filled in, it is overcast. If conditions are completely clear, the circle or triangle is empty. If conditions are partly cloudy, the circle or triangle is partially filled in. The cloud cover shape has different looks depending upon how many oktas (eighths of the sky) are covered by cloud. A sky half full of clouds would have a circle that was half white and half black. Below the shape indicating sky cover, the station model can indicate the coverage of low clouds, in oktas, and the ceiling height in hundreds of feet. The ceiling height is the height at which more than half the sky is covered by clouds.

For pilots, knowledge of the sky cover helps determine if visual flight rules are being met. Knowing the degree of cloud cover can help determine whether or not various weather fronts, such as cold fronts or warm fronts, have passed by a location. A nephanalysis, contouring areas that are cloudy with scalloped lines, can be performed to indicate a system›s cloud and precipitation pattern. This technique is rarely performed nowadays, due to the prevalence of satellite imagery worldwide.

Cloud Types

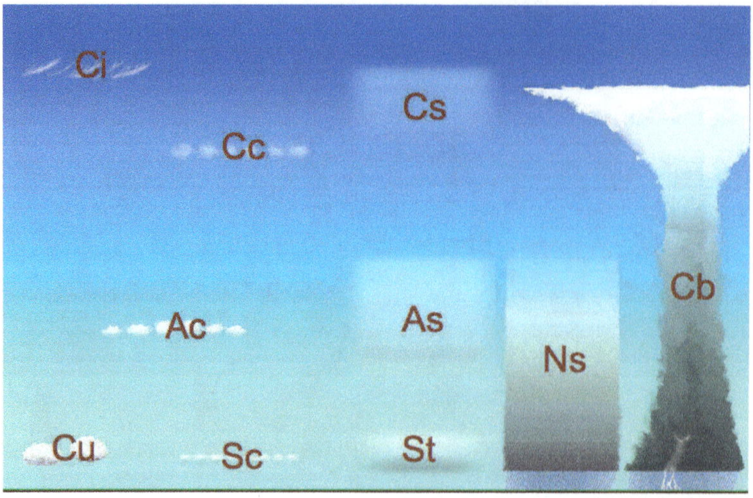

Cloud classification by altitude of occurrence (towering vertical cumulus congestus not shown)

Above or below the circle for manned stations (automatic stations do not report cloud types) that indicates sky cover can lie one or more symbols indicating cloud types in any of the low, middle and high-étages for tropospheric clouds. One predominant cloud type may be depicted for each of three étages, if known. The middle and high-étage types are depicted above the sky cover circle of the station model, while the main low-étage cloud type is indicated below the circle. Since the station model has limited room, it makes no special provision for vertical or multi-level clouds that can occupy more than one étage at a particular time. Consequently, cloud genera with significant vertical development may be coded and plotted as low or middle depending on the altitude at which they normally form. Cumulus and cumulonimbus usually form in the low étage of the troposphere and achieve vertical extent by growing upward into the middle or high étage. Conversely, nimbostratus usually forms in the middle étage of the troposphere and becomes vertically developed by growing downward into the low étage. Although the SYNOP code has no separate formal group classification for vertical or multi-level clouds, the observer procedure for selecting numerical codes is designed to give high reporting priority to those genera or species that show significant vertical development.

The symbols used for clouds emulate the cloud shape. Cirrus is indicated by a couple hooks, cumulus are indicated by a mound shape, with cumulonimbus indicated with an upside down trapezoid on top of the cumulus symbol to indicate its anvil. When there is more than one cloud type present per level, the cloud type with the highest priority is included. Knowing the cloud type at various locations can help determine whether or not a weather front has passed by a particular location. A low deck of stratus could indicate a station is still north of a warm front, while thunderstorms can indicate the approach of a squall line or cold front.

Cloud Abbreviations	Code No.	C_L	Description (Abridged From International Code)
St—STRATUS	1		Cu of fair weather, little vertical development and seemingly flattened
Fra—FRACTUS	2		Cu of considerable development, generally towering, with or without other Cu or Sc bases all at same level
Sc—STRATOCUMULUS	3		Cb with tops lacking clear-cut outlines, but distinctly not cirriform or anvil-shaped; with or without Cu, Sc, or St
Cu—CUMULUS	4		Sc formed by spreading out of Cu; Cu often present also
Cb—CUMULONIMBUS	5		Sc not formed by spreading out of Cu
Ac—ALTOCUMULUS	6		St or StFra, but no StFra of bad weather
Ns—NIMBOSTRATUS	7		StFra and/or CuFra of bad weather (scud)
As—ALTOSTRATUS	8		Cu and Sc (not formed by spreading out of Cu) with bases at different levels
Ci—CIRRUS	9		Cb having a clearly fibrous (cirriform) top, often anvil-shaped, with or without Cu, Sc, St, or scud
Cs—CIRROSTRATUS			
Cc—CIRROCUMULUS			

Low étage (Sc,St) and upward-growing vertical (Cu, Cb)

No.	C_M	Description (Abridged From International Code)
1		Thin As (most of cloud layer semitransparent)
2		Thick As, greater part sufficiently dense to hide sun (or moon), or Ns
3		Thin Ac, mostly semitransparent; cloud elements not changing much and at a single level
4		Thin Ac in patches; cloud elements continually changing and/or occurring at more than one level
5		Thin Ac in bands or in a layer gradually spreading over sky and usually thickening as a whole
6		Ac formed by the spreading out of Cu or Cb
7		Double-layered Ac, or a thick layer of Ac, not increasing; or Ac with As and/or Ns
8		Ac in the form of Cu-shaped tufts or Ac with turrets
9		Ac of a chaotic sky, usually at different levels; patches of dense Ci are usually present also

Middle étage (Ac,As) and downward-growing vertical (Ns)

Code No.	C_H	Description (Abridged From International Code) 3
1		Filaments of Ci, or "mares tails," scattered and not increasing
2		Dense Ci in patches or twisted sheaves, usually not increasing, sometimes like remains of Cb; or towers or tufts
3		Dense Ci, often anvil-shaped, derived from or associated with Cb
4		Ci, often hook-shaped, gradually spreading over the sky and usually thickening as a whole
5		Ci and Cs, often in converging bands, or Cs alone; generally overspreading and growing denser; the continuous layer not reaching 45° altitude
6		Ci and Cs, often in converging bands, or Cs alone; generally overspreading and growing denser; the continuous layer exceeding 45° altitude
7		Veil of Cs covering the entire sky
8		Cs not increasing and not covering entire sky
9		Cc alone or Cc with some Ci or Cs, but the Cc being the main cirriform cloud

High étage (Ci,Cc,Cs)

Present Weather and Visibility

Common present weather symbols

To the left of the cloud shape in the center of the station model is the symbol depicting present weather. The present weather symbol depicts the current weather which normally is obstructing the visibility at the time of observation. The visibility itself is shown as a number, in statute miles in the United States and meters elsewhere, describing how far the observer can see at that time. This number is located to the left of the present weather symbol. For pilots, knowledge of the horizontal visibility helps determine if they are flying through instrument meteorological conditions, such as foggy or smokey conditions, as well as areas of intense precipitation. Present weather depicted with the station model can include:

- Dust
- Fog
- Haze
- Ocean spray
- Precipitation
- Sand
- Smoke
- Thunderstorms
- Volcanic ash

Temperature and Dew Point

To left of center in the station model, the temperature and dew point are plotted. Within the United States on surface weather maps, they are still plotted in degrees Fahrenheit. Otherwise, they will be in units of degrees Celsius. This knowledge is important to meteorologists because when this data is plotted on a map, isotherms and isodrosotherms (lines of equal dew point) are easily analyzed, either by man or machine, which can help determine location of weather fronts.

Sea level Pressure and Height of Pressure Surface

On the top right corner of the model for a surface weather map is the pressure, showing the last two integer digits of the pressure in millibars, or hectopascals, along with the first decimal. For instance, if the pressure at a certain location is 999.7 hPa, the pressure portion of the station model will read 997. Although the first digit or two of the pressure is left off, other nearby stations give away whether the pressure starts with a 10 or a 9. Most of the time, choosing first digits that would lead to a value nearest to 1000 works out best. The plotting of this value within the station model allows for the analysis of isobars on weather maps. Within maps which plot data on constant pressure surfaces, the pressure is replaced with the height of the pressure surface.

Pressure Tendency

Pressure tendency figures

Below the pressure will lie the pressure tendency figure, which indicates the pressure change over the past three hours. The number depicting the pressure change will usually have two digits and indicate the pressure change in 0.1 millibar increments. There are nine different figures which represent the pressure change. A slant angled up and right indicates a steady rise, while a downward slant to the lower right indicates a steady fall. Steady rises can indicate improving conditions and approaching high pressure, and normally occur in the wake of a cold front. Steady falls can indicate deteriorating conditions and an approaching low pressure area, with the largest falls occurring in advance of a surface cyclone and its attendant warm front.

The time of the day must be considered, as there are two natural rises per day (locally around 10 am and 10 pm) and two natural falls per day (locally around 4 am and 4 pm). These daily changes in pressure can mask the movement of pressure systems and fronts past a location. The lowest natural dip in the pressure, in a stagnant weather pattern, occurs around 4 pm while the highest natural peak in pressure occurs around 10 am. Once plotted on a map, the analysis of isallobars (lines of equal pressure change) can be plotted on a map, which can indicate the direction of motion of high and low pressures systems across the map domain.

Code Number	W	past weather
0		Clear or few clouds
1		Partly cloudy
2		Cloudy (broken or overcast)
3	S→ +→	Sandstorm or duststorm OR drifting or blowing snow
4	≡	Fog, smoke, or thick haze
5	,	Drizzle
6	•	Rain
7	*	Snow, rain and snow mixed, or ice pellets
8	▽	Showers
9	ꓘ	Thunderstorm, with or without precipitation

Past weather symbols

Past Weather

Station models can have past weather plotted within them, witch would be located just under the pressure tendency. They indicate the type of weather experienced during the past six hours. The weather types are limited to obstructions of visibility and precipitation.

Plotting of Data

A station model is a symbolic illustration showing the weather occurring at a given reporting station. Meteorologists created the station model to plot a number of weather elements in a small space on weather maps. Maps filled with dense station-model plots can be difficult to read, but they allow meteorologists, pilots, and mariners to see important weather patterns. A computer draws a station model for each observation location. The station model is primarily used on surface-weather maps, but can also be used to show the weather aloft. A completed station-model map allows users to analyze patterns in air pressure, temperature, wind, cloud cover, and precipitation.

Station model plots use an internationally accepted coding convention that has changed little since August 1, 1941. Elements in the plot show the key weather elements, including temperature, dewpoint, wind, cloud cover, air pressure, pressure tendency, and precipitation. Winds have a standard notation when plotted on weather maps. More than a century ago, winds were plotted as arrows, with feathers on just one side depicting five knots of wind, while feathers on both sides depicted 10 knots (19 km/h) of wind. The notation changed to that of half of an arrow, with half of a wind barb indicating five knots, a full barb ten knots, and a pennant flag fifty knots.

Because of the structure of the SYNOP code, a maximum of three cloud symbols can be plotted for each reporting station that appears on the weather map. All cloud types are coded and transmitted by trained observers then plotted on maps as low, middle, or high-étage using special symbols for each major cloud type. Any cloud type with significant vertical extent that can occupy more than one étage is coded as low (cumulus and cumulonimbus) or middle (nimbostratus) depending on the altitude level or étage where it normally initially forms aside from any vertical growth that takes place. The symbol used on the map for each of these étages at a particular observation time is for the genus, species, variety, mutation, or cloud motion that is considered most important according to criteria set out by the World Meteorological Organization (WMO). If these elements for any étage at the time of observation are deemed to be of equal importance, then the type which is predominant in amount is coded by the observer and plotted on the weather map using the appropriate symbol. Special weather maps in aviation show areas of icing and turbulence.

Types

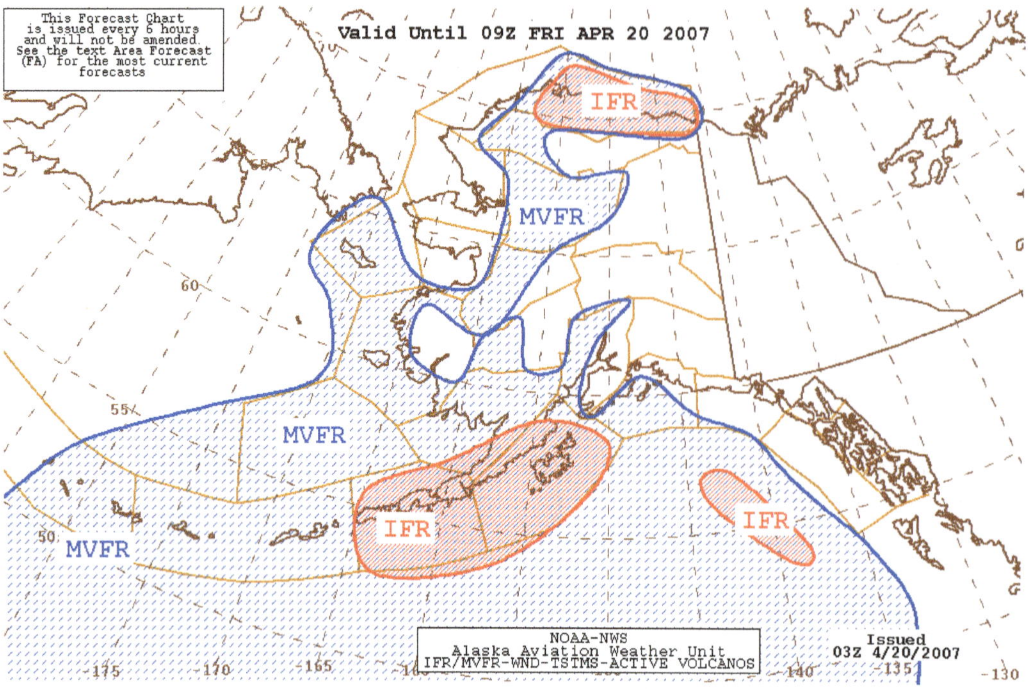

Alaskan aviation weather map

Aviation Maps

Aviation interests have their own set of weather maps. One type of map shows where VFR (visual flight rules) are in effect and where IFR (instrument flight rules) are in effect. Weather depiction plots show ceiling height (level where at least half the sky is covered with clouds) in hundreds of feet, present weather, and cloud cover. Icing maps depict areas where icing can be a hazard for flying. Aviation-related maps also show areas of turbulence.

Constant Pressure Charts

Constant pressure charts normally contain plotted values of temperature, humidity, wind, and the vertical height above sea level of the pressure surface. They have a variety of uses. In the mountainous terrain of the western United States and Mexican Plateau, the 850 hPa pressure surface can be a more realistic depiction of the weather pattern than a standard surface analysis. Using the 850 and 700 hPa pressure surfaces, one can determine when and where warm advection (coincident with upward vertical motion) and cold advection (coincident with downward vertical motion) is occurring within the lower portions of the troposphere. Areas with small dewpoint depressions and are below freezing indicate the presence of icing conditions for aircraft. The 500 hPa pressure surface can be used as a rough guide for the motion of many tropical cyclones. Shallower tropical cyclones, which have experienced vertical wind shear, tend to be steered by winds at the 700 hPa level.

Use of the 300 and 200 hPa constant pressure charts can indicate the strength of systems in the lower troposphere, as stronger systems near the Earth's surface are reflected as stronger features at these levels of the atmosphere. Isotachs are drawn at these levels, which a lines of equal wind speed. They are helpful in finding maxima and minima in the wind pattern. Minima in the wind pattern aloft are favorable for tropical cyclogenesis. Maxima in the wind pattern at various levels of the atmosphere show locations of jet streams. Areas colder than −40 °C (−40 °F) indicate a lack of significant icing, as long as there is no active thunderstorm activity.

Surface Weather Analysis

A surface weather analysis is a type of weather map that depicts positions for high and low-pressure areas, as well as various types of synoptic scale systems such as frontal zones. Isotherms can be drawn on these maps, which are lines of equal temperature. Isotherms are drawn normally as solid lines at a preferred temperature interval. They show temperature gradients, which can be useful in finding fronts, which are on the warm side of large temperature gradients. By plotting the freezing line, isotherms can be useful in determination of precipitation type. Mesoscale boundaries such as tropical cyclones, outflow boundaries and squall lines also are analyzed on surface weather analyses.

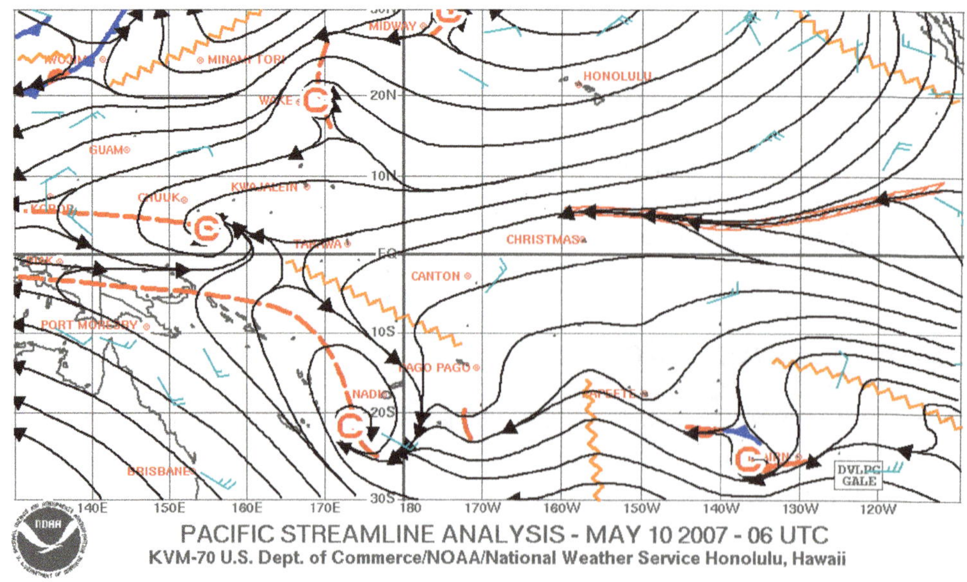

Streamline analysis of the tropical Pacific Ocean

Isobaric analysis is performed on these maps, which involves the construction of lines of equal mean sea level pressure. The innermost closed lines indicate the positions of relative maxima and minima in the pressure field. The minima are called low-pressure areas while the maxima are called high-pressure areas. Highs are often shown as H's whereas lows are shown as L's. Elongated areas of low pressure, or troughs, are sometimes plotted as thick, brown dashed lines down the trough axis. Isobars are commonly used to place surface boundaries from the horse latitudes poleward, while streamline analyses are used in the tropics. A streamline analysis is a series of arrows oriented parallel to wind, showing wind motion within a certain geographic area. C's depict cyclonic flow or likely areas of low pressure, while A's depict anticyclonic flow or likely positions of high-pressure areas. An area of confluent streamlines shows the location of shearlines within the tropics and

REFERENCES

1. Encarta (2009). "Chart". Microsoft Corporation. Retrieved 2007-11-25.
2. DataStreme Atmosphere (2008-04-28). "Air Temperature Patterns". American Meteorological Society. Archived from the original on 2008-05-11. Retrieved 2010-02-07.
3. Jay Snively (2010). "H-I-J". MAPS GPS. Retrieved 2010-01-30.
4. Human Intelligence (2007-07-25). "Francis Galton (1822–1911)". Indiana University. Retrieved 2007-04-18.

5. Allaby, Michael (2009). Atmosphere: A Scientific History of Air, Weather, and Climate. Infobase Publishing. Retrieved 2013-12-07.
6. Steven Roberts. "THE COMPANIES AND THE WEATHER".
7. John D. Cox (2002). Stormwatchers: The Turbulent History of Weather Prediction From Franklin's Kite to El Nino. John Wiley & Sons, Inc. pp. 53–56. ISBN 0-471-38108-X.
8. WebExhibits (2008). "Daylight Saving Time". Idea. Retrieved 2007-06-24.
9. National Oceanic and Atmospheric Administration (2007-05-30). "An Expanding Presence". United States Department of Commerce. Retrieved 2010-01-31.
10. University of Oklahoma. The Norwegian Cyclone Model. Retrieved on 2007-05-17.
11. Bureau of Meteorology (2010). "Air Masses and Weather Maps". Commonwealth of Australia. Retrieved 2010-02-06.
12. Weather Prediction Center (2007-03-01). "A Brief History of the Weather Prediction Center". National Oceanic and Atmospheric Administration. Retrieved 2014-07-01.
13. United States Weather Bureau (1948-07-01). "Daily Weather Map". United States Department of Commerce. Retrieved 2010-02-06.
14. United States Weather Bureau (1954-05-14). "Daily Weather Map". United States Department of Commerce. Retrieved 2010-02-06.
15. National Meteorological Center (January 1969). "Prospectus for an NMC Digital Facsimile Incoder Mapping Program" (PDF). Environmental Science Services Administration. Retrieved 2007-05-05.
16. Hong Kong Observatory (2009-09-03). "The Hong Kong Observatory Computer System and Its Applications". The Government of the Hong Kong Special Administrative Region. Retrieved 2010-02-06.
17. Hydrometeorological Prediction Center (2000). "Hydrometeorological Prediction Center 1999 Accomplishment Report". National Oceanic and Atmospheric Administration. Retrieved 2007-05-05.
18. David M. Roth (2006-12-14). "Unified Surface Analysis Manual" (PDF). Hydrometeorological Prediction Center. Retrieved 2006-10-22.
19. S. A. Saseendran, L. Harenduprakash, L. S. Rathore and V. S. Singh (2004-12-05). "A GIS application for weather analysis and forecasting". GISDevelopment.net. Retrieved 2007-05-05.
20. CoCoRAHS (2005). Introduction to Drawing Isopleths. Colorado Climate Center. Retrieved on 2007-04-29.
21. National Weather Service (2003). Station Model Example. Retrieved on 2007-04-29.
22. Dr Elizabeth R. Tuttle (2005). Weather Maps. J. B. Calvert. Retrieved on 2007-05-10.

23. World Meteorological Organization, ed. (1975). Étages, International Cloud Atlas (PDF). I. pp. 15–16. ISBN 92-63-10407-7. Retrieved 26 August 2014.
24. Koermer, Jim (2011). «Plymouth State Meteorology Program Cloud Boutique». Plymouth State University.
25. NOAA, ed. (September 3, 2007). Federal Meteorological Handbook (FMH) Number 2 (PDF). NOAA. p. C-17. Retrieved 26 November 2014.
26. Unisys Corportation (2009). «Surface Data Details». Retrieved 2010-02-07.
27. Jeppesen (2008-05-06). «Weather Help». Archived from the original on 2008-06-07. Retrieved 2010-02-07.
28. Glossary of Meteorology (June 2000). «Constant-pressure chart». American Meteorological Society. Retrieved 2010-02-06.
29. Terry T. Lankford (1999). Aircraft icing: a pilot›s guide. McGraw-Hill Professional. pp. 129–134. ISBN 0-07-134139-0. Retrieved 2010-02-06.
30. United States Navy (2007). «Section 1. Influences on Tropical Cyclone Motion». Retrieved 2010-02-06.
31. Edward J. Hopkins, Ph.D. (1996-06-10). «Surface Weather Analysis Chart». University of Wisconsin. Retrieved 2007-05-10.
32. Bureau of Meteorology (2010). «The Weather Map». Commonwealth of Australia. Retrieved 2010-02-06.
33. National Weather Service Forecast Office Honolulu, Hawaii (2010-02-07). «Pacific Streamline Analysis». Pacific Region Headquarters. Retrieved 2010-02-07.
34. CoCoRAHS (2015).INTRODUCTION TO DRAWING ISOPLETHS. Colorado Climate Center. Retrieved on 2007-04-29.
35. National Weather Service (2003). Station Model Example. Retrieved on 2007-04-29.
36. Dr Elizabeth R. Tuttle (2005). Weather Maps. J. B. Calvert. Retrieved on 2007-05-10.
37. Hydrometeorological Prediction Center (2008). Decoding the station model. National Centers for Environmental Prediction. Retrieved on 2007-05-16.
38. JetStream (2008). How to read weather maps. National Weather Service. Retrieved on 2007-05-16.
39. Terry T. Lankford (2000). Aviation Weather Handbook. McGraw-Hill Professional. ISBN 978-0-07-136103-3. Retrieved on 2008-01-22.
40. United States Weather Bureau (1871). Daily Weather Map: January 1, 1871. Retrieved on 2008-01-22.
41. United States Weather Bureau (1941). Daily Weather Map: July 31, 1941. Retrieved on 2008-01-22.
42. United States Weather Bureau (1941). Daily Weather Map: August 1, 1941. Retrieved on 2008-01-22.
43. MetOffice (2008). Interpreting weather charts. Retrieved on 2007-05-17.

44. Glossary of Meteorology (2009). Nephanalysis. American Meteorological Society. Retrieved on 2008-01-22.
45. AEROGRAPHER›S MATE 1&C (2008). SHORT-RANGE EXTRAPOLATION. Integrated Publishing. Retrieved on 2008-01-22.
46. NOAA, ed. (September 3, 2007). Federal Meteorological Handbook (FMH) Number 2 (PDF). NOAA. p. C-17. Retrieved 26 November 2014.
47. Clouds Online (2012). «Cloud Atlas». Retrieved 1 February 2012.
48. JetStream (2003). Cloud type priority via Internet Wayback Machine. National Weather Service Southern Region Headquarters. Retrieved on 2007-05-17.
49. American Meteorological Society (2006). Map Symbols. Retrieved on 2008-01-22.
50. University of Wisconsin–Madison Department of Atmospheric and Ocean Sciences (1996). Air Masses and Fronts. Retrieved on 2008-01-22.
51. Frank Singleton (2000). Tides in the Atmosphere. Retrieved on 2007-05-16.
52. Dr P. M. Inness. Isobaric analysis and pressure tendency. Dr. Giles Harrison. Retrieved on 2008-01-22.

CHAPTER 5

Meteorology

Chapter Contents

Meteorology is an extremely interdisciplinary science, drawing on the laws of physics and chemistry (among others) to aid in our understanding of Earth's atmosphere, its processes, and its structure. It is a study that dates to ancient times, when ancient civilizations made observations and kept records of weather conditions, both for agricultural purposes and out of a general curiosity about the world around them. Over the centuries, the atmosphere has been studied for a variety of reasons, including agricultural knowledge, military defense and planning, and developing better warnings for severe weather systems like tornadoes and hurricanes. Technological advances, such as the development of scientific computing and an increase in the total number of meteorological observations being taken daily across the globe, have allowed for better forecasts (or at least the meteorological community likes to think they are better forecasts) and a much better overall understanding of our atmosphere.

INTRODUCTION

Meteorology is the interdisciplinary scientific study of the atmosphere. The study of meteorology dates back millennia, though significant progress in meteorology did not occur until the 18th century. The 19th century saw modest progress in the field after weather observation networks were formed across broad regions. Prior attempts at prediction of weather depended on historical data. It wasn't until after the elucidation of the laws of physics and, more particularly, the development of the computer, allowing for the automated solution of a great many equations that model the weather, in the latter half of the 20th century that significant breakthroughs in weather forecasting were achieved.

Meteorological phenomena are observable weather events that are explained by the science of meteorology. Meteorological phenomena are described and quantified by the variables of Earth's atmosphere: temperature, air pressure, water vapor, mass flow, and the variations and interactions of those variables, and how they change over time. Different spatial scales are used to describe and predict weather on local, regional, and global levels.

Meteorology, climatology, atmospheric physics, and atmospheric chemistry are sub-disciplines of the atmospheric sciences. Meteorology and hydrology compose the interdisciplinary field of hydrometeorology. The interactions between Earth's atmosphere and its oceans are part of a coupled ocean-atmosphere system. Meteorology has application in many diverse fields such as the military, energy production, transport, agriculture, and construction.

The word "meteorology" is from Greek μετέωρος metéōros lofty; high (in the sky) and -λογία -logia "-(o)logy, i.e. the study of things in the air.

The beginnings of meteorology can be traced back to ancient India, as the Upanishads contain serious discussion about the processes of cloud formation and rain and the seasonal cycles caused by the movement of Earth around the sun. Varāhamihira›s classical work *Brihatsamhita*, written about 500 AD, provides clear evidence that a deep knowledge of atmospheric processes existed even in those times.

In 350 BC, Aristotle wrote *Meteorology*. Aristotle is considered the founder of meteorology. One of the most impressive achievements described in the *Meteorology* is the description of what is now known as the hydrologic cycle.

The book De Mundo (composed before 250 BC or between 350 and 200 BC) noted

If the flashing body is set on fire and rushes violently to the Earth it is called a thunderbolt; if it is only half of fire, but violent also and massive, it is called a *meteor*; if it is entirely free from fire, it is called a smoking bolt. They are all called 'swooping bolts' because they swoop down upon the Earth. Lightning is sometimes smoky, and is then called 'smoldering lightning"; sometimes it darts quickly along, and is then said to be *vivid*. At other times, it travels in crooked lines, and is called *forked lightning*. When it swoops down upon some object it is called 'swooping lightning'.

The Greek scientist Theophrastus compiled a book on weather forecasting, called the *Book of Signs*. The work of Theophrastus remained a dominant influence in the study of weather and in weather forecasting for nearly 2,000 years. In 25 AD, Pomponius Mela, a geographer for the Roman Empire, formalized the climatic zone system. According to Toufic Fahd, around the 9th century, Al-Dinawari wrote the *Kitab al-Nabat* (*Book of*

Plants), in which he deals with the application of meteorology to agriculture during the Muslim Agricultural Revolution. He describes the meteorological character of the sky, the planets and constellations, the sun and moon, the lunar phases indicating seasons and rain, the *anwa* (heavenly bodies of rain), and atmospheric phenomena such as winds, thunder, lightning, snow, floods, valleys, rivers, lakes.

Research of Visual Atmospheric Phenomena

Twilight at Baker Beach

Ptolemy wrote on the atmospheric refraction of light in the context of astronomical observations. In 1021, Alhazen showed that atmospheric refraction is also responsible for twilight; he estimated that twilight begins when the sun is 19 degrees below the horizon, and also used a geometric determination based on this to estimate the maximum possible height of the Earth's atmosphere as 52,000 *passim* (about 49 miles, or 79 km).

St. Albert the Great was the first to propose that each drop of falling rain had the form of a small sphere, and that this form meant that the rainbow was produced by light interacting with each raindrop. Roger Bacon was the first to calculate the angular size of the rainbow. He stated that a rainbow summit can not appear higher than 42 degrees above the horizon. In the late 13th century and early 14th century, Kamāl al-Dīn al-Fārisī and Theodoric of Freiberg were the first to give the correct explanations for the primary rainbow phenomenon. Theoderic went further and also explained the secondary rainbow. In 1716, Edmund Halley suggested that aurorae are caused by magnetic effluvia moving along the Earth›s magnetic field lines.

Instruments and Classification Scales

THE ROBINSON ANEMOMETER.

A hemispherical cup anemometer

In 1441, King Sejong's son, Prince Munjong, invented the first standardized rain gauge. These were sent throughout the Joseon Dynasty of Korea as an official tool to assess land taxes based upon a farmer's potential harvest. In 1450, Leone Battista Alberti developed a swinging-plate anemometer, and was known as the first *anemometer*. In 1607, Galileo Galilei constructed a thermoscope. In 1611, Johannes Kepler wrote the first scientific treatise on snow crystals: "Strena Seu de Nive Sexangula (A New Year's Gift of Hexagonal Snow)". In 1643, Evangelista Torricelli invented the mercury barometer. In 1662, Sir Christopher Wren invented the mechanical, self-emptying, tipping bucket rain gauge. In 1714, Gabriel Fahrenheit created a reliable scale for measuring temperature with a mercury-type thermometer. In 1742, Anders Celsius, a Swedish astronomer, proposed the "centigrade" temperature scale, the predecessor of the current Celsius scale. In 1783, the first hair hygrometer was demonstrated by Horace-Bénédict de Saussure. In 1802–1803, Luke Howard wrote *On the Modification of Clouds*, in which he assigns cloud types Latin names. In 1806, Francis Beaufort introduced his system for classifying wind speeds. Near the end of the 19th century the first cloud atlases were published, including the *International Cloud Atlas*, which has remained in print ever since. The April 1960 launch of the first successful weather satellite, TIROS-1, marked the beginning of the age where weather information became available globally.

Atmospheric Composition Research

In 1648, Blaise Pascal rediscovered that atmospheric pressure decreases with height, and deduced that there is a vacuum above the atmosphere. In 1738, Daniel Bernoulli published *Hydrodynamics*, initiating the Kinetic theory of gases and established the basic laws for the theory of gases. In 1761, Joseph Black discovered that ice absorbs heat without changing its temperature when melting. In 1772, Black's student Daniel Rutherford discovered nitrogen, which he called *phlogisticated air*, and together they developed the phlogiston theory. In 1777, Antoine Lavoisier discovered oxygen and developed an explanation for combustion. In 1783, in Lavoisier's essay "Reflexions sur le phlogistique", he deprecates the phlogiston theory and proposes a caloric theory. In 1804, Sir John Leslie observed that a matte black surface radiates heat more effectively than a polished surface, suggesting the importance of black body radiation. In 1808, John Dalton defended caloric theory in *A New System of Chemistry* and described how it combines with matter, especially gases; he proposed that the heat capacity of gases varies inversely with atomic weight. In 1824, Sadi Carnot analyzed the efficiency of steam engines using caloric theory; he developed the notion of a reversible process and, in postulating that no such thing exists in nature, laid the foundation for the second law of thermodynamics.

Research Into Cyclones and Air Flow

In 1494, Christopher Columbus experienced a tropical cyclone, which led to the first written European account of a hurricane. In 1686, Edmund Halley presented a systematic study of the trade winds and monsoons and identified solar heating as the cause of atmospheric motions. In 1735, an *ideal* explanation of global circulation through study of the trade winds was written by George Hadley. In 1743, when Benjamin Franklin was prevented from seeing a lunar eclipse by a hurricane, he decided that cyclones move in a contrary manner to the winds at their periphery. Understanding the kinematics of how exactly the rotation of the Earth affects airflow was partial at first. Gaspard-Gustave Coriolis published a paper in 1835 on the energy yield of machines with rotating parts, such as waterwheels. In 1856, William Ferrel proposed the existence of a circulation cell in the mid-latitudes, and the air within deflected by the Coriolis force resulting in the prevailing westerly winds. Late in the 19th century, the motion of air masses along isobars was understood to be the result of the large-scale interaction of the pressure gradient force and the deflecting force. By 1912, this deflecting force was named the Coriolis effect. Just after World War I, a group of meteorologists in Norway led by Vilhelm Bjerknes developed the Norwegian cyclone model that explains the generation, intensification and ultimate decay (the life cycle) of mid-latitude cyclones, and introduced the idea of fronts, that is, sharply defined boundaries between air masses. The group included Carl-Gustaf Rossby (who was the first to explain the large scale atmospheric flow in terms of fluid dynamics), Tor Bergeron (who first determined how rain forms) and Jacob Bjerknes.

Observation Networks and Weather Forecasting

In 1654, Ferdinando II de Medici established the first *weather observing* network, that consisted of meteorological stations in Florence, Cutigliano, Vallombrosa, Bologna, Parma,

Milan, Innsbruck, Osnabrück, Paris and Warsaw. The collected data were sent to Florence at regular time intervals. In 1832, an electromagnetic telegraph was created by Baron Schilling. The arrival of the electrical telegraph in 1837 afforded, for the first time, a practical method for quickly gathering surface weather observations from a wide area. This data could be used to produce maps of the state of the atmosphere for a region near the Earth›s surface and to study how these states evolved through time. To make frequent weather forecasts based on these data required a reliable network of observations, but it was not until 1849 that the Smithsonian Institution began to establish an observation network across the United States under the leadership of Joseph Henry. Similar observation networks were established in Europe at this time. The Reverend William Clement Ley was key in understanding of cirrus clouds and early understandings of Jet Streams. Later after this Charles Kenneth Mackinnon Douglas known as ‹CKM› Douglas read Ley›s papers after his death and carried on the early study of weather systems. Nineteenth century researchers in meteorology were drawn from military or medical backgrounds, rather than trained as dedicated scientists. In 1854, the United Kingdom government appointed Robert FitzRoy to the new office of *Meteorological Statist to the Board of Trade* with the task of gathering weather observations at sea. FitzRoy›s office became the United Kingdom Meteorological Office in 1854, the first national meteorological service in the world. The first daily weather forecasts made by FitzRoy›s Office were published in *The Times* newspaper in 1860. The following year a system was introduced of hoisting storm warning cones at principal ports when a gale was expected.

Over the next 50 years many countries established national meteorological services. The India Meteorological Department (1875) was established to follow tropical cyclone and monsoon. The Finnish Meteorological Central Office (1881) was formed from part of Magnetic Observatory of Helsinki University. Japan›s Tokyo Meteorological Observatory, the forerunner of the Japan Meteorological Agency, began constructing surface weather maps in 1883. The United States Weather Bureau (1890) was established under the United States Department of Agriculture. The Australian Bureau of Meteorology (1906) was established by a Meteorology Act to unify existing state meteorological services.

NUMERICAL WEATHER PREDICTION

Numerical weather prediction (NWP) uses mathematical models of the atmosphere and oceans to predict the weather based on current weather conditions. Though first attempted in the 1920s, it was not until the advent of computer simulation in the 1950s that numerical weather predictions produced realistic results. A number of global and regional forecast models are run in different countries worldwide, using current weather observations relayed from radiosondes, weather satellites and other observing systems as inputs.

Mathematical models based on the same physical principles can be used to generate either short-term weather forecasts or longer-term climate predictions; the latter are widely applied for understanding and projecting climate change. The improvements made to regional models have allowed for significant improvements in tropical cyclone track and air quality forecasts; however, atmospheric models perform poorly at handling processes that occur in a relatively constricted area, such as wildfires.

Manipulating the vast datasets and performing the complex calculations necessary to modern numerical weather prediction requires some of the most powerful supercomputers in the world. Even with the increasing power of supercomputers, the forecast skill of numerical weather models extends to only about six days. Factors affecting the accuracy of numerical predictions include the density and quality of observations used as input to the forecasts, along with deficiencies in the numerical models themselves. Post-processing techniques such as model output statistics (MOS) have been developed to improve the handling of errors in numerical predictions.

A more fundamental problem lies in the chaotic nature of the partial differential equations that govern the atmosphere. It is impossible to solve these equations exactly, and small errors grow with time (doubling about every five days). Present understanding is that this chaotic behavior limits accurate forecasts to about 14 days even with perfectly accurate input data and a flawless model. In addition, the partial differential equations used in the model need to be supplemented with parameterizations for solar radiation, moist processes (clouds and precipitation), heat exchange, soil, vegetation, surface water, and the effects of terrain. In an effort to quantify the large amount of inherent uncertainty remaining in numerical predictions, ensemble forecasts have been used since the 1990s to help gauge the confidence in the forecast, and to obtain useful results farther into the future than otherwise possible. This approach analyzes multiple forecasts created with an individual forecast model or multiple models.

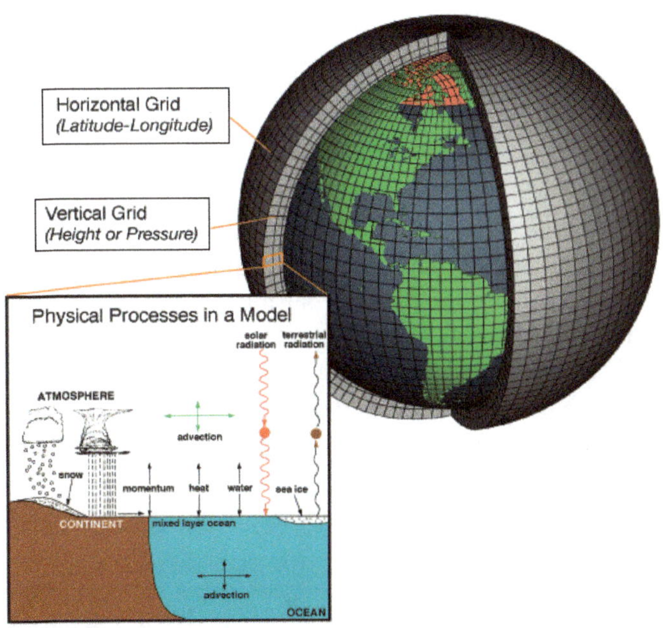

Weather models use systems of differential equations based on the laws of physics, fluid motion, and chemistry, and use a coordinate system which divides the planet into a 3D grid. Winds, heat transfer, solar radiation, relative humidity, and surface hydrology are calculated within each grid cell, and the interactions with neighboring cells are used to calculate atmospheric properties in the future.

The history of numerical weather prediction began in the 1920s through the efforts of Lewis Fry Richardson, who used procedures originally developed by Vilhelm Bjerknes to produce by hand a six-hour forecast for the state of the atmosphere over two points in central Europe, taking at least six weeks to do so. It was not until the advent of the computer and computer simulations that computation time was reduced to less than the forecast period itself. The ENIAC was used to create the first weather forecasts via computer in 1950, based on a highly simplified approximation to the atmospheric governing equations. In 1954, Carl-Gustav Rossby›s group at the Swedish Meteorological and Hydrological Institute used the same model to produce the first operational forecast (i.e., a routine prediction for practical use). Operational numerical weather prediction in the United States began in 1955 under the Joint Numerical Weather Prediction Unit (JNWPU), a joint project by the U.S. Air Force, Navy and Weather Bureau. In 1956, Norman Phillips developed a mathematical model which could realistically depict monthly and seasonal patterns in the troposphere; this became the first successful climate model. Following Phillips› work, several groups began working to create general circulation models. The first general circulation climate model that combined both oceanic and atmospheric processes was developed in the late 1960s at the NOAAGeophysical Fluid Dynamics Laboratory.

As computers have become more powerful, the size of the initial data sets has increased and newer atmospheric models have been developed to take advantage of the added available computing power. These newer models include more physical processes in the simplifications of the equations of motion in numerical simulations of the atmosphere. In 1966, West Germany and the United States began producing operational forecasts based on primitive-equation models, followed by the United Kingdom in 1972 and Australia in 1977. The development of limited area (regional) models facilitated advances in forecasting the tracks of tropical cyclones as well as air quality in the 1970s and 1980s. By the early 1980s models began to include the interactions of soil and vegetation with the atmosphere, which led to more realistic forecasts.

The output of forecast models based on atmospheric dynamics is unable to resolve some details of the weather near the Earth›s surface. As such, a statistical relationship between the output of a numerical weather model and the ensuing conditions at the ground was developed in the 1970s and 1980s, known as model output statistics (MOS). Starting in the 1990s, model ensemble forecasts have been used to help define the forecast uncertainty and to extend the window in which numerical weather forecasting is viable farther into the future than otherwise possible.

Initialization

The atmosphere is a fluid. As such, the idea of numerical weather prediction is to sample the state of the fluid at a given time and use the equations of fluid dynamics and thermodynamics to estimate the state of the fluid at some time in the future. The process of entering observation data into the model to generate initial conditions is called *initialization*. On land, terrain maps available at resolutions down to 1 kilometer (0.6 mi) globally are used to help model atmospheric circulations within regions of rugged topography, in order to better depict features such as downslope winds, mountain waves and related cloudiness that affects incoming solar radiation. The main inputs from country-based weather services are observations from devices (called radiosondes) in weather balloons that measure various atmospheric parameters and transmits them to a fixed receiver, as well as from weather satellites. The World Meteorological Organization acts to standardize the instrumentation, observing practices and timing of these observations worldwide. Stations either report hourly in METAR reports, or every six hours in SYNOP reports. These observations are irregularly spaced, so they are processed by data assimilation and objective analysis methods, which perform quality control and obtain values at locations usable by the model's mathematical algorithms. The data are then used in the model as the starting point for a forecast.

Weather reconnaissance aircraft, such as this WP-3D Orion, provide data that is then used in numerical weather forecasts.

A variety of methods are used to gather observational data for use in numerical models. Sites launch radiosondes in weather balloons which rise through the troposphere and well into the stratosphere. Information from weather satellites is used where traditional data sources are not available. Commerce provides pilot reports along aircraft routes and ship reports along shipping routes. Research projects use reconnaissance aircraft to fly in and around weather systems of interest, such as tropical cyclones. Reconnaissance aircraft are also flown over the open oceans during the cold season into systems which cause significant uncertainty in forecast guidance, or are expected to be of high impact from three to seven days into the future over the downstream continent. Sea ice began to be initialized in forecast models in 1971. Efforts to involve sea surface temperature in model initialization began in 1972 due to its role in modulating weather in higher latitudes of the Pacific.

Computation

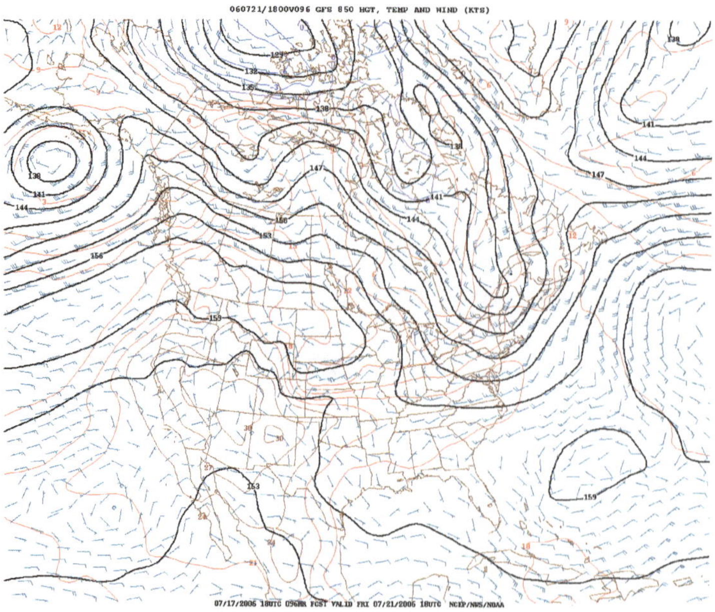

A prognostic chart of the 96-hour forecast of 850 mbar geopotential height and temperature from the Global Forecast System

An atmospheric model is a computer program that produces meteorological information for future times at given locations and altitudes. Within any modern model is a set of equations, known as the primitive equations, used to predict the future state of the atmosphere.These equations—along with the ideal gas law—are used to evolve the density, pressure, and potential temperature scalar fields and the air velocity (wind) vector field of the atmosphere through time. Additional transport equations for pollutants and other aerosols are included in some primitive-equation high-resolution models as well. The equations used are nonlinear partial differential equations which are impossible to solve exactly through analytical methods, with the exception of a few idealized cases. Therefore, numerical methods obtain approximate solutions. Different models use different solution methods: some global models and almost all regional models use finite difference methods for all three spatial dimensions, while other global models and a few regional models use spectral methods for the horizontal dimensions and finite-difference methods in the vertical.

These equations are initialized from the analysis data and rates of change are determined. These rates of change predict the state of the atmosphere a short time into the future; the time increment for this prediction is called a *time step*. This future atmospheric state is then used as the starting point for another application of the predictive equations to find new rates of change, and these new rates of change predict the atmosphere at a yet further time step into the future. This time stepping is repeated until the solution reaches the desired forecast time. The length of the time step chosen within the model is related to the distance between the points on the computational grid, and is chosen to maintain numerical stability. Time steps for global models are on the order of tens of minutes, while time steps for regional models are between one and four minutes. The global models are run at varying times into the future. The UKMET Unified Model is run six days into the future, while the European Centre for Medium-Range Weather Forecasts' Integrated Forecast System and Environment Canada's Global Environmental Multiscale Model both run out to ten days into the future, and the Global Forecast System model run by the Environmental Modeling Center is run sixteen days into the future. The visual output produced by a model solution is known as a prognostic chart, or *prog*.

Parameterization

Field of cumulus clouds, which are parameterized since they are too small to be explicitly included within numerical weather prediction

Some meteorological processes are too small-scale or too complex to be explicitly included in numerical weather prediction models. *Parameterization* is a procedure for representing these processes by relating them to variables on the scales that the model resolves. For example, the gridboxes in weather and climate models have sides that are between 5 kilometers (3 mi) and 300 kilometers (200 mi) in length. A typical cumulus cloud has a scale of less than 1 kilometer (0.6 mi), and would require a grid even finer than this to be represented physically by the equations of fluid motion. Therefore, the processes that such clouds represent are parameterized, by processes of various sophistication. In the earliest models, if a column of air within a model gridbox was conditionally unstable (essentially, the bottom was warmer and moister than the top) and the water vapor content at any point within the column became saturated then it would be overturned (the warm, moist air would begin rising), and the air in that vertical column mixed. More sophisticated schemes recognize that only some portions of the box might convect and that entrainment and other processes occur. Weather models that have gridboxes with sides between 5 and 25 kilometers (3 and 16 mi) can explicitly represent convective clouds, although they need to parameterize cloud microphysics which occur at a smaller scale. The formation of large-scale (stratus-type) clouds is more physically based; they form when the relative humidity reaches some prescribed value. Sub-grid scale processes need to be taken into account. Rather than assuming that clouds form at 100% relative humidity, the cloud fraction can be related to a critical value of relative humidity less than 100%, reflecting the sub grid scale variation that occurs in the real world.

The amount of solar radiation reaching the ground, as well as the formation of cloud droplets occur on the molecular scale, and so they must be parameterized before they can be included in the model. Atmospheric drag produced by mountains must also be parameterized, as the limitations in the resolution of elevation contours produce significant underestimates of the drag. This method of parameterization is also done for the surface flux of energy between the ocean and the atmosphere, in order to determine realistic sea surface temperatures and type of sea ice found near the ocean›s surface. Sun angle as well as the impact of multiple cloud layers is taken into account. Soil type, vegetation type, and soil moisture all determine how much radiation goes into warming and how much moisture is drawn up into the adjacent atmosphere, and thus it is important to parameterize their contribution to these processes. Within air quality models, parameterizations take into account atmospheric emissions from multiple relatively tiny sources (e.g. roads, fields, factories) within specific grid boxes.

Domains

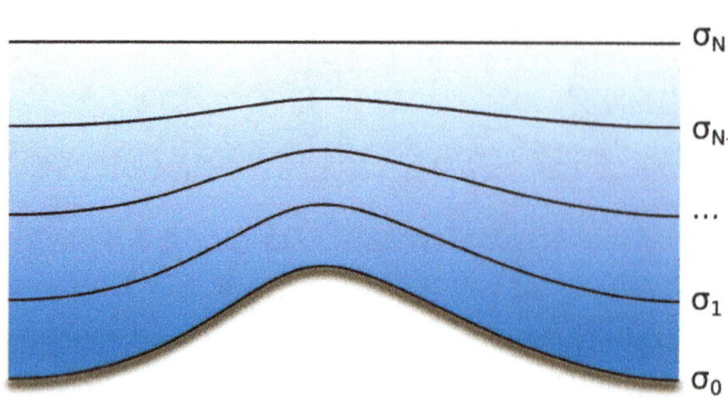

A cross-section of the atmosphere over terrain with a sigma-coordinate representation shown. Mesoscale models divide the atmosphere vertically using representations similar to the one shown here.

The horizontal domain of a model is either *global*, covering the entire Earth, or *regional*, covering only part of the Earth. Regional models (also known as *limited-area* models, or LAMs) allow for the use of finer grid spacing than global models because the available computational resources are focused on a specific area instead of being spread over the globe. This allows regional models to resolve explicitly smaller-scale meteorological phenomena that cannot be represented on the coarser grid of a global model. Regional models use a global model to specify conditions at the edge of their domain (boundary conditions) in order to allow systems from outside the regional model domain to move into its area. Uncertainty and errors within regional models are introduced by the global model used for the boundary conditions of the edge of the regional model, as well as errors attributable to the regional model itself.

Coordinate Systems

Horizontal Coordinates

Horizontal position may be expressed directly in geographic coordinates (latitude and longitude) for global models or in a map projectionplanar coordinates for regional models.

Vertical Coordinates

The vertical coordinate is handled in various ways. Lewis Fry Richardson's 1922 model used geometric height z) as the vertical coordinate. Later models substituted the geometric z coordinate with a pressure coordinate system, in which the geopotential heights of constant-pressure surfaces become dependent variables, greatly simplifying the primitive equations.This correlation between coordinate systems can be made since pressure decreases with height through the Earth›s atmosphere. The first model used for operational forecasts, the single-layer barotropic model, used a single pressure coordinate at the 500-millibar (about 5,500 m (18,000 ft)) level, and thus was essentially two-dimensional. High-resolution models—also called *mesoscale models*—such as the Weather Research and Forecasting model tend to use normalized pressure coordinates referred to as sigma coordinates. This coordinate system receives its name from the independent variable σ used to scale atmospheric pressures with respect to the pressure at the surface, and in some cases also with the pressure at the top of the domain.

Model Output Statistics

Because forecast models based upon the equations for atmospheric dynamics do not perfectly determine weather conditions, statistical methods have been developed to attempt to correct the forecasts. Statistical models were created based upon the three-dimensional fields produced by numerical weather models, surface observations and the climatological conditions for specific locations. These statistical models are collectively referred to as model output statistics (MOS), and were developed by the National Weather Service for their suite of weather forecasting models in the late 1960s.

Model output statistics differ from the *perfect prog* technique, which assumes that the output of numerical weather prediction guidance is perfect. MOS can correct for local effects that cannot be resolved by the model due to insufficient grid resolution, as well as model biases. Because MOS is run after its respective global or regional model, its production is known as post-processing. Forecast parameters within MOS include maximum and minimum temperatures, percentage chance of rain within a several hour period, precipitation amount expected, chance that the precipitation will be frozen in nature, chance for thunderstorms, cloudiness, and surface winds.

Ensembles

In 1963, Edward Lorenz discovered the chaotic nature of the fluid dynamics equations involved in weather forecasting. Extremely small errors in temperature, winds, or other initial inputs given to numerical models will amplify and double every five days, making it impossible for long-range forecasts—those made more than two weeks in advance—to predict the state of the atmosphere with any degree of forecast skill. Furthermore, existing observation networks have poor coverage in some regions (for example, over large bodies of water such as the Pacific Ocean), which introduces uncertainty into the true initial state of the atmosphere. While a set of equations, known as the Liouville equations, exists to determine the initial uncertainty in the model initialization, the equations are too complex to run in real-time, even with the use of supercomputers. These uncertainties limit forecast model accuracy to about five or six days into the future.

Edward Epstein recognized in 1969 that the atmosphere could not be completely described with a single forecast run due to inherent uncertainty, and proposed using an ensemble of stochastic Monte Carlo simulations to produce means and variances for the state of the atmosphere. Although this early example of an ensemble showed skill, in 1974 Cecil Leith showed that they produced adequate forecasts only when the ensemble probability distribution was a representative sample of the probability distribution in the atmosphere.

Since the 1990s, *ensemble forecasts* have been used operationally (as routine forecasts) to account for the stochastic nature of weather processes – that is, to resolve their inherent uncertainty. This method involves analyzing multiple forecasts created with an individual forecast model by using different physical parametrizations or varying initial conditions. Starting in 1992 with ensemble forecasts prepared by the European Centre for Medium-Range Weather Forecasts (ECMWF) and the National Centers for Environmental Prediction, model ensemble forecasts have been used to help define the forecast uncertainty and to extend the window in which numerical weather forecasting is viable farther into the future than otherwise possible. The ECMWF model, the Ensemble Prediction System,uses singular vectors to simulate the initial probability density, while the NCEP ensemble, the Global Ensemble Forecasting System, uses a technique known as vector breeding. The UK Met Office runs global and regional ensemble forecasts where perturbations to initial conditions are produced using a Kalman filter. There are 24 ensemble members in the Met Office Global and Regional Ensemble Prediction System (MOGREPS).

In a single model-based approach, the ensemble forecast is usually evaluated in terms of an average of the individual forecasts concerning one forecast variable, as well as the degree of agreement between various forecasts within the ensemble system, as represented by their overall spread. Ensemble spread is diagnosed through tools such as spaghetti diagrams, which show the dispersion of one quantity on prognostic charts for specific time steps in the future. Another tool where ensemble spread is used is a meteogram, which shows the dispersion in the forecast of one quantity for one specific location. It is common for the ensemble spread to be too small to include the weather that actually occurs, which can lead to forecasters misdiagnosing model uncertainty; this problem becomes particularly severe for forecasts of the weather about ten days in advance. When ensemble spread is small and the forecast solutions are consistent within multiple model runs, forecasters perceive more confidence in the ensemble mean, and the forecast in general. Despite this perception, a *spread-skill relationship* is often weak or not found, as spread-error correlations are normally less than 0.6, and only under special circumstances range between 0.6–0.7. The relationship between ensemble spread and forecast skill varies substantially depending on such factors as the forecast model and the region for which the forecast is made.

In the same way that many forecasts from a single model can be used to form an ensemble, multiple models may also be combined to produce an ensemble forecast. This approach is called *multi-model ensemble forecasting*, and it has been shown to improve forecasts when compared to a single model-based approach. Models within a multi-model ensemble can be adjusted for their various biases, which is a process known as *superensemble forecasting*. This type of forecast significantly reduces errors in model output.

Applications

Air Quality Modeling

Air quality forecasting attempts to predict when the concentrations of pollutants will attain levels that are hazardous to public health. The concentration of pollutants in the atmosphere is determined by their *transport*, or mean velocity of movement through the atmosphere, their diffusion, chemical transformation, and ground deposition. In addition to pollutant source and terrain information, these models require data about the state of the fluid flow in the atmosphere to determine its transport and diffusion. Meteorological conditions such as thermal inversions can prevent surface air from rising, trapping pollutants near the surface, which makes accurate forecasts of such events crucial for air quality modeling. Urban air quality models require a very fine computational mesh, requiring the use of high-resolution mesoscale weather models; in spite of this, the quality of numerical weather guidance is the main uncertainty in air quality forecasts.

Climate Modeling

A General Circulation Model (GCM) is a mathematical model that can be used in computer simulations of the global circulation of a planetary atmosphere or ocean. An atmospheric general circulation model (AGCM) is essentially the same as a global numerical weather prediction model, and some (such as the one used in the UK Unified Model) can be configured for both short-term weather forecasts and longer-term climate predictions. Along with sea ice and land-surface components, AGCMs and oceanic GCMs (OGCM) are key components of global climate models, and are widely applied for understanding the climate and projecting climate change. For aspects of climate change, a range of man-made chemical emission scenarios can be fed into the climate models to see how an enhanced greenhouse effect would modify the Earth's climate. Versions designed for climate applications with time scales of decades to centuries were originally created in 1969 by Syukuro Manabe and Kirk Bryan at the Geophysical Fluid Dynamics Laboratory in Princeton, New Jersey. When run for multiple decades, computational limitations mean that the models must use a coarse grid that leaves smaller-scale interactions unresolved.

Ocean Surface Modeling

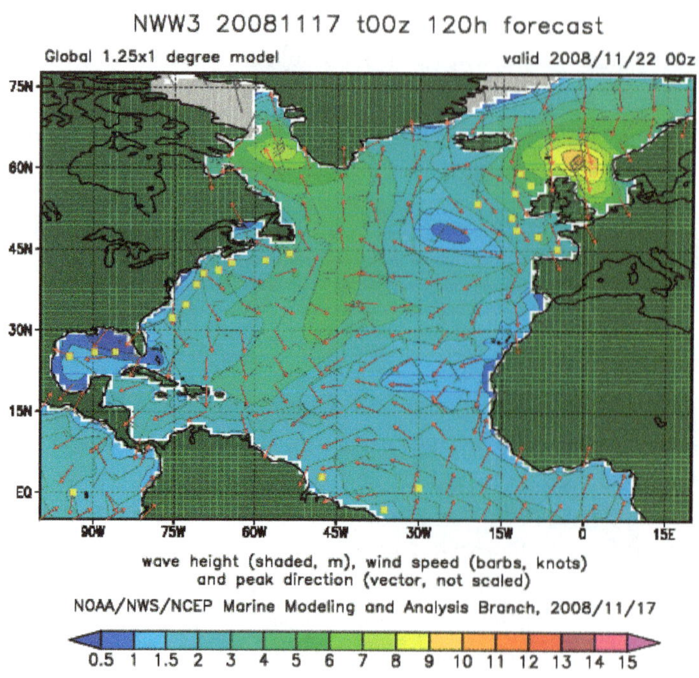

NOAA Wavewatch III 120-hour wind and wave forecast for the North Atlantic

The transfer of energy between the wind blowing over the surface of an ocean and the ocean's upper layer is an important element in wave dynamics. The spectral wave transport equation is used to describe the change in wave spectrum over changing topography. It simulates wave generation, wave movement (propagation within a fluid), wave shoaling, refraction, energy transfer between waves, and wave dissipation. Since surface winds are the primary forcing mechanism in the spectral wave transport equation, ocean wave models use information produced by numerical weather prediction models as inputs to determine how much energy is transferred from the atmosphere into the layer at the surface of the ocean. Along with dissipation of energy through whitecaps and resonance between waves, surface winds from numerical weather models allow for more accurate predictions of the state of the sea surface.

Tropical Cyclone Forecasting

Tropical cyclone forecasting also relies on data provided by numerical weather models. Three main classes of tropical cyclone guidance models exist: Statistical models are based on an analysis of storm behavior using climatology, and correlate a storm's position and date to produce a forecast that is not based on the physics of the atmosphere at the time. Dynamical models are numerical models that solve the governing equations of fluid flow in the atmosphere; they are based on the same principles as other limited-area numerical weather prediction models but may include special computational techniques such as refined spatial domains that move along with the cyclone. Models that use elements of both approaches are called statistical-dynamical models.

In 1978, the first hurricane-tracking model based on atmospheric dynamics—the movable fine-mesh (MFM) model—began operating. Within the field of tropical cyclone track forecasting, despite the ever-improving dynamical model guidance which occurred with increased computational power, it was not until the 1980s when numerical weather prediction showed skill, and until the 1990s when it consistently outperformed statistical or simple dynamical models. Predictions of the intensity of a tropical cyclone based on numerical weather prediction continue to be a challenge, since statistical methods continue to show higher skill over dynamical guidance.

Wildfire Modeling

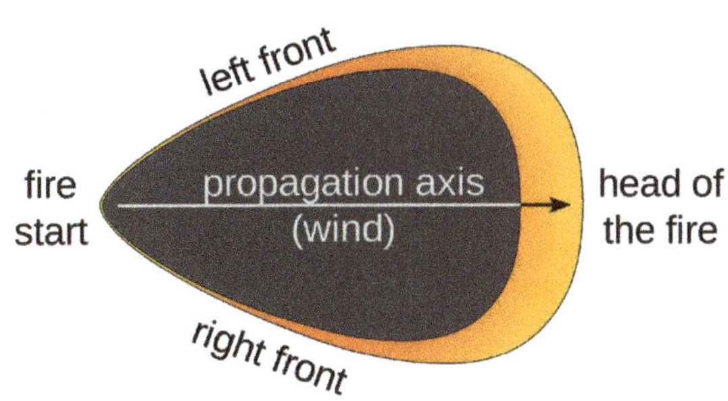

A simple wildfire propagation model

On a molecular scale, there are two main competing reaction processes involved in the degradation of cellulose, or wood fuels, in wildfires. When there is a low amount of moisture in a cellulose fiber, volatilization of the fuel occurs; this process will generate intermediate gaseous products that will ultimately be the source of combustion. When moisture is present—or when enough heat is being carried away from the fiber, charring occurs. The chemical kinetics of both reactions indicate that there is a point at which the level of moisture is low enough—and/or heating rates high enough—for combustion processes become self-sufficient. Consequently, changes in wind speed, direction, moisture, temperature, or lapse rate at different levels of the atmosphere can have a significant impact on the behavior and growth of a wildfire. Since the wildfire acts as a heat source to the atmospheric flow, the wildfire can modify local advection patterns, introducing a feedback loop between the fire and the atmosphere.

A simplified two-dimensional model for the spread of wildfires that used convection to represent the effects of wind and terrain, as well as radiative heat transfer as the dominant method of heat transport led to reaction-diffusion systems of partial differential equations. More complex models join numerical weather models or computational fluid dynamics models with a wildfire component which allow the feedback effects between the fire and the atmosphere to be estimated. The additional complexity in the latter class of models translates to a corresponding increase in their computer power requirements. In fact, a full three-dimensional treatment of combustion via direct numerical simulation at scales relevant for atmospheric modeling is not currently practical because of the excessive computational cost such a simulation would require. Numerical weather models have limited forecast skill at spatial resolutions under 1 kilometer (0.6 mi), forcing complex wildfire models to parameterize the fire in order to calculate how the winds will be modified locally by the wildfire, and to use those modified winds to determine the rate at which the fire will spread locally. Although models such as Los Alamos' FIRETEC solve for the concentrations of fuel and oxygen, the computational grid cannot be fine enough to resolve the combustion reaction, so approximations must be made for the temperature distribution within each grid cell, as well as for the combustion reaction rates themselves.

Meteorologists

Meteorologists are scientists who study meteorology. The American Meteorological Society published and continually updates an authoritative electronic *Meteorology Glossary*. Meteorologists work in government agencies, private consulting and research services, industrial enterprises, utilities, radio and television stations, and in education. In the United States, meteorologists held about 9,400 jobs in 2009.

Meteorologists are best known by the public for weather forecasting. Some radio and television weather forecasters are professional meteorologists, while others are reporters (weather specialist, weatherman, etc.) with no formal meteorological training. The American Meteorological Society and National Weather Association issue "Seals of Approval" to weather broadcasters who meet certain requirements.

Equipment

Each science has its own unique sets of laboratory equipment. In the atmosphere, there are many things or qualities of the atmosphere that can be measured. Rain, which can be observed, or seen anywhere and anytime was one of the first atmospheric qualities measured historically. Also, two other accurately measured qualities are wind and humidity. Neither of these can be seen but can be felt. The devices to measure these three sprang up in the mid-15th century and were respectively the rain gauge, the anemometer, and the hygrometer. Many attempts had been made prior to the 15th century to construct adequate equipment to measure the many atmospheric variables. Many were faulty in some way or were simply not reliable. Even Aristotle noted this in some of his work; as the difficulty to measure the air.

Sets of surface measurements are important data to meteorologists. They give a snapshot of a variety of weather conditions at one single location and are usually at a weather station, a ship or a weather buoy. The measurements taken at a weather station can include any number of atmospheric observables. Usually, temperature, pressure, wind measurements, and humidity are the variables that are measured by a thermometer, barometer, anemometer, and hygrometer, respectively. Professional stations may also include air quality sensors (carbon monoxide, carbon dioxide, methane, ozone, dust, and smoke), ceilometer (cloud ceiling), falling precipitation sensor, flood sensor, lightning sensor, microphone (explosions, sonic booms, thunder), pyranometer/pyrheliometer/spectroradiometer (IR/Vis/UV photodiodes), rain gauge/snow gauge, scintillation counter (background radiation, fallout, radon), seismometer (earthquakes and tremors), transmissometer (visibility), and a GPS clock for data logging. Upper air data are of crucial importance for weather forecasting. The most widely used technique is launches of radiosondes. Supplementing the radiosondes a network of aircraft collection is organized by the World Meteorological Organization.

Satellite image of Hurricane Hugo with a polar low visible at the top of the image.

Remote sensing, as used in meteorology, is the concept of collecting data from remote weather events and subsequently producing weather information. The common types of remote sensing are Radar, Lidar, and satellites (or photogrammetry). Each collects data about the atmosphere from a remote location and, usually, stores the data where the instrument is located. Radar and Lidar are not passive because both use EM radiation to illuminate a specific portion of the atmosphere. Weather satellites along with more general-purpose Earth-observing satellites circling the earth at various altitudes have become an indispensable tool for studying a wide range of phenomena from forest fires to El Niño.

Spatial Scales

The study of the atmosphere can be divided into distinct areas that depend on both time and spatial scales. At one extreme of this scale is climatology. In the timescales of hours to days, meteorology separates into micro-, meso-, and synoptic scale meteorology. Respectively, the geospatial size of each of these three scales relates directly with the appropriate timescale.

Other subclassifications are used to describe the unique, local, or broad effects within those subclasses.

Typical Scales of Atmospheric Motion Systems

Type of motion	Horizontal scale (meter)
Molecular mean free path	10^{-3}
Minute turbulent eddies	10^{-2} - 10^{-1}
Small eddies	10^{-1} - 1
Dust devils	1 - 10
Gusts	10 - 10^{2}
Tornadoes	10^{2}
Thunderclouds	10^{3}
Fronts, squall lines	10^{4} - 10^{5}
Hurricanes	10^{5}
Synoptic Cyclones	10^{6}
Planetary waves	10^{7}
Atmospheric tides	10^{7}
Mean zonal wind	10^{7}

Microscale

Microscale meteorology is the study of atmospheric phenomena on a scale of about 1 kilometre (0.62 mi) or less. Individual thunderstorms, clouds, and local turbulence caused by buildings and other obstacles (such as individual hills) are modeled on this scale.

Mesoscale

Mesoscale meteorology is the study of atmospheric phenomena that has horizontal scales ranging from 1 km to 1000 km and a vertical scale that starts at the Earth's surface and includes the atmospheric boundary layer, troposphere, tropopause, and the lower section of the stratosphere. Mesoscale timescales last from less than a day to weeks. The events typically of interest are thunderstorms, squall lines, fronts, precipitation bands in tropical and extratropical cyclones, and topographically generated weather systems such as mountain waves and sea and land breezes.

SYNOPTIC SCALE METEOROLOGY

The synoptic scale in meteorology (also known as large scale or cyclonic scale) is a horizontal length scale of the order of 1000 kilometres (about 620 miles) or more. This corresponds to a horizontal scale typical of mid-latitude depressions (e.g. extratropical cyclones). Most high and low-pressure areas seen on weather maps such as surface weather analyses are synoptic-scale systems, driven by the location of Rossby waves in their respective hemisphere. Low-pressure areas and their related frontal zones occur on the leading edge of a trough within the Rossby wave pattern, while surface highs form on the back edge of the trough. Most precipitation areas occur near frontal zones. The word synoptic is derived from the Greek word συνοπτικός (synoptikos), meaning seen together.

The Navier–Stokes equations applied to atmospheric motion can be simplified by scale analysis in the synoptic scale. It can be shown that the main terms in horizontal equations are Coriolis force and pressure gradient terms; therefore, one can use geostrophic approximation. In vertical coordinates, the momentum equation simplifies to the hydrostatic equilibrium equation.

Extratropical Cyclone

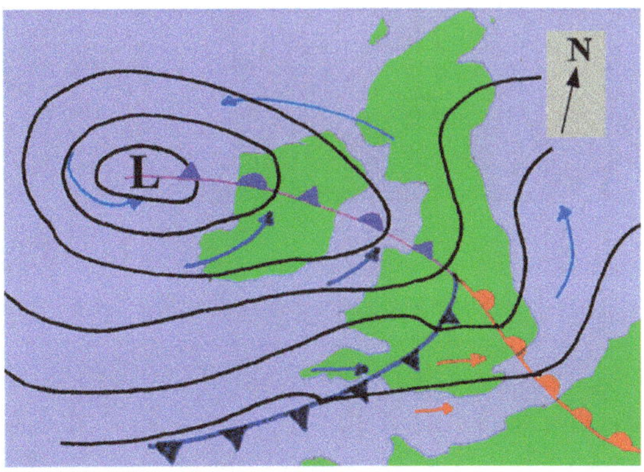

A fictitious synoptic chart of an extratropical cyclone affecting the UK and Ireland. The blue arrows between isobars indicate the direction of the wind, while the "L" symbol denotes the centre of the "low". Note the occluded, cold and warm frontal boundaries.

An extratropical cyclone is a synoptic scale low-pressure weather system that has neither tropical nor polar characteristics, being connected with fronts and horizontal gradients in temperature and dew point otherwise known as "baroclinic zones.

The descriptor extratropical refers to the fact that this type of cyclone generally occurs outside of the tropics, in the middle latitudes of the planet. These systems may also be described as mid-latitude cyclones due to their area of formation, or post-tropical cyclones where extratropical transition has occurred, and are often described as depressions or lows by weather forecasters and the public. These are the everyday phenomena which along with anti-cyclones, drive the weather over much of the Earth.

Although extratropical cyclones are almost always classified as baroclinic since they form along zones of temperature and dewpoint gradient within the westerlies, they can sometimes become barotropic late in their life cycle when the temperature distribution around the cyclone becomes fairly uniform with radius. An extratropical cyclone can transform into a subtropical storm, and from there into a tropical cyclone, if it dwells over warm waters and develops central convection, which warms its core.

Surface High-Pressure Systems

Golden Gate Bridge in fog

High-pressure systems are frequently associated with light winds at the surface and subsidence through the lower portion of the troposphere. Subsidence will generally dry out an air mass by adiabatic, or compressional, heating. Thus, high pressure typically brings clear skies.During the day, since no clouds are present to reflect sunlight, there is more incoming shortwave solar radiation and temperatures rise. At night, the absence of clouds means that outgoing longwave radiation (i.e. heat energy from the surface) is not absorbed, giving cooler diurnal low temperatures in all seasons. When surface winds become light, the subsidence produced directly under a high-pressure system can lead to a buildup of particulates in urban areas under the ridge, leading to widespread haze. If the low level relative humidity rises towards 100 percent overnight, fog can form.

Strong, vertically shallow high-pressure systems moving from higher latitudes to lower latitudes in the northern hemisphere are associated with continental arctic air masses. The low, sharp inversion can lead to areas of persistent stratocumulus or stratus cloud, colloquially known as anticyclonic gloom. The type of weather brought about by an anticyclone depends on its origin. For example, extensions of the Azores high pressure may bring about anticyclonic gloom during the winter, as they are warmed at the base and will trap moisture as they move over the warmer oceans. High pressures that build to the north and extend southwards will often bring clear weather. This is due to being cooled at the base (as opposed to warmed) which helps prevent clouds from forming.

On weather maps, these areas show converging winds (isotachs), also known as confluence, or converging height lines near or above the level of non-divergence, which is near the 500 hPa pressure surface about midway up through the troposphere. High-pressure systems are alternatively referred to as anticyclones. On weather maps, high-pressure centers are associated with the letter H in English, or A in Spanish, because alta is the Spanish word for high, within the isobar with the highest pressure value. On constant pressure upper level charts, it is located within the highest height line contour.

Weather Fronts

A weather front is a boundary separating two masses of air of different densities, and is the principal cause of meteorological phenomena. In surface weather analyses, fronts are depicted using various colored lines and symbols, depending on the type of front. The air masses separated by a front usually differ in temperature and humidity. Cold fronts may feature narrow bands of thunderstorms and severe weather, and may on occasion be preceded by squall lines or dry lines. Warm fronts are usually preceded by stratiform precipitation and fog. The weather usually clears quickly after a front's passage. Some fronts produce no precipitation and little cloudiness, although there is invariably a wind shift.

Cold fronts and occluded fronts generally move from west to east, while warm fronts move poleward. Because of the greater density of air in their wake, cold fronts and cold occlusions move faster than warm fronts and warm occlusions. Mountains and warm bodies of water can slow the movement of fronts. When a front becomes stationary, and the density contrast across the frontal boundary vanishes, the front can degenerate into a line which separates regions of differing wind velocity, known as a shearline. This is most common over the open ocean.

Global Scale

Global scale meteorology is the study of weather patterns related to the transport of heat from the tropics to the poles. Very large scale oscillations are of importance at this scale. These oscillations have time periods typically on the order of months, such as the Madden–Julian oscillation, or years, such as the El Niño–Southern Oscillation and the Pacific decadal oscillation. Global scale meteorology pushes into the range of climatology. The traditional definition of climate is pushed into larger timescales and with the understanding of the longer time scale global oscillations, their effect on climate and weather disturbances can be included in the synoptic and mesoscale timescales predictions.

Numerical Weather Prediction is a main focus in understanding air–sea interaction, tropical meteorology, atmospheric predictability, and tropospheric/stratospheric processes. The Naval Research Laboratory in Monterey, California, developed a global atmospheric model called Navy Operational Global Atmospheric Prediction System (NOGAPS). NOGAPS is run operationally at Fleet Numerical Meteorology and Oceanography Center for the United States Military. Many other global atmospheric models are run by national meteorological agencies.

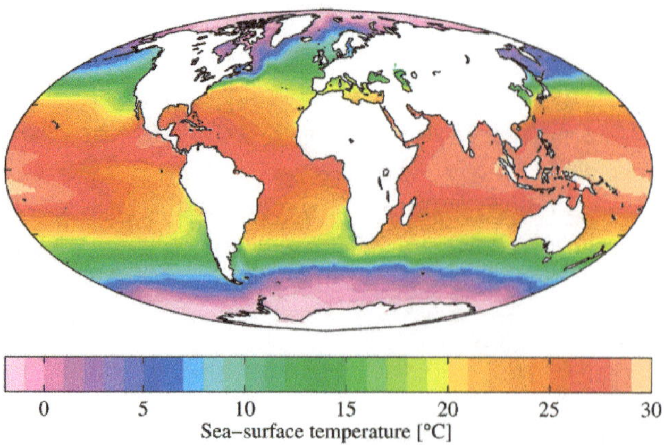

Annual mean sea surface temperatures.

Some Meteorological Principles

Boundary Layer Meteorology

Boundary layer meteorology is the study of processes in the air layer directly above Earth's surface, known as the atmospheric boundary layer (ABL). The effects of the surface – heating, cooling, and friction – cause turbulent mixing within the air layer. Significant movement of heat, matter, or momentum on time scales of less than a day are caused by turbulent motions. Boundary layer meteorology includes the study of all types of surface–atmosphere boundary, including ocean, lake, urban land and non-urban land for the study of meteorology.

Dynamic Meteorology

Dynamic meteorology generally focuses on the fluid dynamics of the atmosphere. The idea of air parcel is used to define the smallest element of the atmosphere, while ignoring the discrete molecular and chemical nature of the atmosphere. An air parcel is defined as a point in the fluid continuum of the atmosphere. The fundamental laws of fluid dynamics, thermodynamics, and motion are used to study the atmosphere. The physical quantities that characterize the state of the atmosphere are temperature, density, pressure, etc. These variables have unique values in the continuum.

Applications

Weather Forecasting

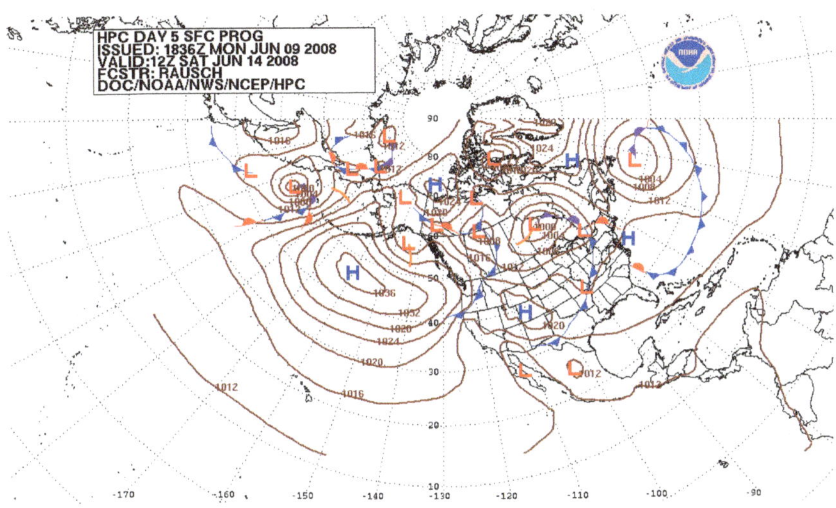

Forecast of surface pressures five days into the future for the north Pacific, North America, and north Atlantic Ocean

Weather forecasting is the application of science and technology to predict the state of the atmosphere at a future time and given location. Humans have attempted to predict the weather informally for millennia and formally since at least the 19th century. Weather forecasts are made by collecting quantitative data about the current state of the atmosphere and using scientific understanding of atmospheric processes to project how the atmosphere will evolve.

Once an all-human endeavor based mainly upon changes in barometric pressure, current weather conditions, and sky condition,forecast models are now used to determine future conditions. Human input is still required to pick the best possible forecast model to base the forecast upon, which involves pattern recognition skills, teleconnections, knowledge of model performance, and knowledge of model biases. The chaotic nature of the atmosphere, the massive computational power required to solve the equations that describe the atmosphere, error involved in measuring the initial conditions, and an incomplete understanding of atmospheric processes mean that forecasts become less accurate as the difference in current time and the time for which the forecast is being made (the *range* of the forecast) increases. The use of ensembles and model consensus help narrow the error and pick the most likely outcome.

There are a variety of end uses to weather forecasts. Weather warnings are important forecasts because they are used to protect life and property. Forecasts based on temperature and precipitation are important to agriculture, and therefore to commodity traders within stock markets. Temperature forecasts are used by utility companies to estimate demand over coming days. On an everyday basis, people use weather forecasts to determine what to wear on a given day. Since outdoor activities are severely curtailed by heavy rain, snow and the wind chill, forecasts can be used to plan activities around these events, and to plan ahead and survive them.

Aviation Meteorology

Aviation meteorology deals with the impact of weather on air traffic management. It is important for air crews to understand the implications of weather on their flight plan as well as their aircraft, as noted by the Aeronautical Information Manual:

The effects of ice on aircraft are cumulative—thrust is reduced, drag increases, lift lessens, and weight increases. The results are an increase in stall speed and a deterioration of aircraft performance. In extreme cases, 2 to 3 inches of ice can form on the leading edge of the airfoil in less than 5 minutes. It takes but 1/2 inch of ice to reduce the lifting power of some aircraft by 50 percent and increases the frictional drag by an equal percentage.

Agricultural Meteorology

Meteorologists, soil scientists, agricultural hydrologists, and agronomists are persons concerned with studying the effects of weather and climate on plant distribution, crop yield, water-use efficiency, phenology of plant and animal development, and the energy balance of managed and natural ecosystems. Conversely, they are interested in the role of vegetation on climate and weather.

Hydrometeorology

Hydrometeorology is the branch of meteorology that deals with the hydrologic cycle, the water budget, and the rainfall statistics of storms. A hydrometeorologist prepares and issues forecasts of accumulating (quantitative) precipitation, heavy rain, heavy snow, and highlights areas with the potential for flash flooding. Typically the range of knowledge that is required overlaps with climatology, mesoscale and synoptic meteorology, and other geosciences.

The multidisciplinary nature of the branch can result in technical challenges, since tools and solutions from each of the individual disciplines involved may behave slightly differently, be optimized for different hard- and software platforms and use different data formats. There are some initiatives - such as the DRIHM project - that are trying to address this issue.

Nuclear Meteorology

Nuclear meteorology investigates the distribution of radioactive aerosols and gases in the atmosphere.

Maritime Meteorology

Maritime meteorology deals with air and wave forecasts for ships operating at sea. Organizations such as the Ocean Prediction Center, Honolulu National Weather Service forecast office, United Kingdom Met Office, and JMA prepare high seas forecasts for the world's oceans.

Military Meteorology

Military meteorology is the research and application of meteorology for military purposes. In the United States, the United States Navy's Commander, Naval Meteorology and Oceanography Command oversees meteorological efforts for the Navy and Marine Corps while the United States Air Force's Air Force Weather Agency is responsible for the Air Force and Army.

Environmental Meteorology

Environmental meteorology mainly analyzes industrial pollution dispersion physically and chemically based on meteorological parameters such as temperature, humidity, wind, and various weather conditions.

Renewable Energy

Meteorology applications in renewable energy includes basic research, "exploration", and potential mapping of wind power and solar radiation for wind and solar energy.

REFERENCES

1. NS, nsimd@ymail.com. "History of Meteorology in India". Imd.gov.in. Retrieved 2012-03-25.
2. "meteorology: Introduction". Retrieved 16 June 2015.
3. "94.05.01: Meteorology". Retrieved 16 June 2015.

4. Aristotle (2004) 350 BCE. Meteorology. The University of Adelaide Library, University of Adelaide, South Australia 5005: eBooks@Adelaide. Translated by E.W. Webster
5. Aristotle; Forster, E. S. (Edward Seymour), 1879-1950; Dobson, J. F. (John Frederic), 1875-1947 (1914). De Mundo. p. Chapter 4.
6. "Weather". Retrieved 16 June 2015.
7. "Timeline of geography, paleontology". Paleorama.com. Following the path of Discovery
8. Fahd, Toufic. "Botany and agriculture": 815.
9. Morelon, Régis; Rashed, Roshdi (1996). Encyclopedia of the History of Arabic Science. 3. Routledge. ISBN 0-415-12410-7.
10. Smith AM, 1996. "Ptolemy's Theory of Visual Perception: An English Translation of the Optics," pp. 46. Transactions of the American Philosophical Society vol. 86, part 2.
11. Frisinger H (1973), "Aristotle's Legacy in Meteorology." Bulletin of the American Meteorological Society volume 3 issue 3, pp. 198–204. Link.
12. "Ancient and pre-Renaissance Contributors to Meteorology". Retrieved 16 June 2015.
13. The rainbow bridge: rainbows in art, myth, and science, p.156, Raymond L. Lee, Alistair B. Fraser
14. Topdemir, Hüseyin Gazi (2007), Kamal Al-din Al-Farisi's explanation of the rainbow
15. Jacobson, Mark Z. (June 2005). Fundamentals of Atmospheric Modeling (paperback) (2nd ed.). New York: Cambridge University Press. p. 828. ISBN 978-0-521-54865-6.
16. "Early Snow Crystal Observations". Retrieved 16 June 2015.
17. Grigull, U., Fahrenheit, a Pioneer of Exact Thermometry. Heat Transfer, 1966, The Proceedings of the 8th International Heat Transfer Conference, San Francisco, 1966, Vol. 1.
18. Beckman, Olof,History of the Celsius temperature scale., translated, Anders Celsius (Elementa,84:4,2001); English
19. Thornes, John. E. (1999). John Constable's Skies. The University of Birmingham Press, pp. 189. ISBN 1-902459-02-4.
20. Bill Giles O.B.E. (2009). Beaufort Scale. BBC. Retrieved on 2009-05-12.
21. Florin to Pascal, September 1647, Œuves completes de Pascal, 2:682.
22. O›Connor, John J.; Robertson, Edmund F., Meteorology, MacTutor History of Mathematics archive, University of St Andrews.
23. Biographical note at Lectures and Papers of Professor Daniel Rutherford (1749–1819), and Diary of Mrs Harriet Rutherford.
24. Sur la combustion en général (On Combustion in general, 1777) and Considérations

Générales sur la Nature des Acides (General Considerations on the Nature of Acids, 1778).

25. Nicholas W. Best, Lavoisier›s ‹Reflections on Phlogiston› I: Against Phlogiston Theory, Foundations of Chemistry, 2015, 17, 137-151.
26. Nicholas W. Best, Lavoisier›s ‹Reflections on Phlogiston› II: On the Nature of Heat, Foundations of Chemistry, 2015, 17. In this early work, Lavoisier calls it "igneous fluid".
27. The 1880 edition of A Guide to the Scientific Knowledge of Things Familiar, a 19th-century educational science book, explained heat transfer in terms of the flow of caloric.
28. Morison, Samuel Eliot, Admiral of the Ocean Sea: A Life of Cristopher Columbus, Boston, 1942, page 617.
29. Cook, Alan H., Edmond Halley: Charting the Heavens and the Seas (Oxford: Clarendon Press, 1998)
30. George Hadley, Concerning the cause of the general trade winds, Philosophical Transactions, vol. 39 (1735).
31. Dorst, Neal, FAQ:_Hurricanes,_Typhoons,_and_Tropical_Cyclones:_Hurricane_Timeline, Hurricane_Research_Division,_Atlantic_Oceanographic_and_Meteorological_Laboratory,_NOAA, January 2006.
32. G-G Coriolis (1835). Sur les équations du mouvement relatif des systèmes de corps. J. De l›Ecole royale polytechnique. 15: 144–154.
33. William Ferrel. An Essay on the Winds and the Currents of the Ocean. Retrieved on 2009-01-01.
34. Arthur Gordon Webster (1912). The Dynamics of Particles and of Rigid, Elastic, and Fluid Bodies. B.G. Teubner. p. 320.
35. Shaye Johnson. The Norwegian Cyclone Model. Retrieved on 2006-10-11.
36. Raymond S. Bradley, Philip D. Jones, Climate Since A.D. 1500, Routledge, 1992, ISBN 0-415-07593-9, p.144
37. Rebecca Martin (2009) Catapult - Indepth - Communication: telegraph, Australian Broadcasting Corporation, Retrieved on 2009-05-12
38. Library of Congress, The Invention of the Telegraph, Retrieved on 2009-01-01
39. Smithsonian Institution Archives. Smithsonian Institution Archives. Retrieved 16 June 2015.
40. Prophet without Honour: The Reverend William Clement Ley and the hunt for the jet stream. rmets.org. Retrieved 13 October 2016.
41. Field, M. (1 October 1999). Meteorologist›s profile — Charles Kenneth Mackinnon Douglas, OBE, AFC, MA. 54 (10): 321–327. doi:10.1002/j.1477-8696.1999.tb03992.x. Retrieved 13 October 2016 – via Wiley Online Library.
42. Williamson, Fiona (2015-09-01). Weathering the empire: meteorological research in the early British straits settlements. The British Journal for the History of Science. 48 (3): 475–492. doi:10.1017/S000708741500028X. ISSN 1474-001X.

43. India Meteorological Department Establishment of IMD. Retrieved on 2009-01-01.
44. Finnish Meteorological Institute. History of Finnish Meteorological Institute. Retrieved on 2009-01-01.
45. Japan Meteorological Agency. History. Retrieved on 2006-10-22.
46. BOM celebrates 100 years. Australian Broadcasting Corporation. Retrieved 2008-01-01.
47. Collections in Perth: 20. Meteorology. National Archives of Australia. Retrieved 2008-05-24.
48. Berknes, V. (1904) Das Problem der Wettervorhersage, betrachtet vom Standpunkte der Mechanik und der Physik (The problem of weather prediction, considered from the viewpoints of mechanics and physics), Meteorologische Zeitschrift, 21 : 1–7. Available in English on-line at: Schweizerbart science publishers.
49. Pioneers in Modern Meteorology and Climatology: Vilhelm and Jacob Bjerknes (PDF). Retrieved 2008-10-13.
50. Richardson, Lewis Fry, Weather Prediction by Numerical Process (Cambridge, England: Cambridge University Press, 1922). Available on-line at: Internet Archive.org.
51. American Institute of Physics. Atmospheric General Circulation Modeling. Retrieved on 2008-01-13.
52. Cox, John D. (2002). Storm Watchers. John Wiley & Sons, Inc. p. 208. ISBN 0-471-38108-X.
53. Edward N. Lorenz, Deterministic non-periodic flow, Journal of the Atmospheric Sciences, vol. 20, pages 130–141 (1963).
54. Manousos, Peter (2006-07-19). Ensemble Prediction Systems. Hydrometeorological Prediction Center. Retrieved 2010-12-31.
55. Glickman, Todd S. (June 2009). Meteorology Glossary (electronic) (2nd ed.). Cambridge, Massachusetts: American Meteorological Society. Retrieved March 10, 2014.
56. Glickman, Todd S. (June 2000). Meteorology Glossary (electronic) (2nd ed.). Cambridge, Massachusetts: American Meteorological Society. Retrieved March 10, 2014.
57. Bureau of labor statistics: Occupational Outlook Handbook, 2010-11 Edition
58. Office of the Federal Coordinator of Meteorology. Federal Meteorological Handbook No. 1 - Surface Weather Observations and Reports: September 2005. Retrieved on 2009-01-02.
59. Peebles, Peyton, 1998, Radar Principles, John Wiley & Sons, Inc., New York, ISBN 0-471-25205-0.
60. Holton, James. An Introduction to Dynamic Meteorology (PDF). Elsevier Academic Press. p. 5. Retrieved 5 March 2016.
61. AMS Glossary of Meteorology. American Meteorological Society. Retrieved 2008-04-12.

62. Online Glossary of Meteorology, American Meteorological Society , 2nd Ed., 2000, Allen Press.
63. Bluestein, H., Synoptic-Dynamic Meteorology in Midlatitudes: Principles of Kinematics and Dynamics, Vol. 1, Oxford University Press, 1992; ISBN 0-19-506267-1
64. Global Modelling, US Naval Research Laboratory, Monterey, Ca.
65. Garratt, J.R., The atmospheric boundary layer, Cambridge University Press, 1992; ISBN 0-521-38052-9.
66. Holton, J.R. 2004. An Introduction to Dynamic Meteorology, 4th Ed., Burlington, Md: Elsevier Inc.. ISBN 0-12-354015-1.
67. Mistic House, Astrology Lessons, History, Prediction, Skeptics, and Astrology Compatibility, Retrieved on 2008-01-12, Archived at: Archived June 8, 2008, at the Wayback Machine.
68. Eric D. Craft, An Economic History of Weather Forecasting, Retrieved on 2007-04-15
69. NASA, Weather Forecasting Through the Ages, Retrieved on 2008-05-25
70. Weather Doctor, Applying The Barometer To Weather Watching, Retrieved on 2008-05-25
71. Mark Moore, Field Forecasting - A Short Summary, Retrieved on 2008-05-25
72. Klaus Weickmann, Jeff Whitaker, Andres Roubicek and Catherine Smith. The Use of Ensemble Forecasts to Produce Improved Medium Range (3-15 days) Weather Forecasts. Retrieved on 2007-02-16.
73. Todd Kimberlain. Tropical cyclone motion and intensity talk (June 2007). Retrieved on 2007-07-21.
74. Richard J. Pasch, Mike Fiorino, and Chris Landsea. TPC/NHC'S REVIEW OF THE NCEP PRODUCTION SUITE FOR 2006. Retrieved on 2008-05-05.
75. National Weather Service Mission Statement, Retrieved on 2008-05-25
76. Blair Fannin, Dry weather conditions continue for Texas, Retrieved on 2008-05-26
77. Dr. Terry Mader, Drought Corn Silage, Retrieved on 2008-05-26
78. Kathryn C. Taylor, Peach Orchard Establishment and Young Tree Care, Retrieved on 2008-05-26
79. Associated Press, After Freeze, Counting Losses to Orange Crop, Retrieved on 2008-05-26
80. The New York Times, FUTURES/OPTIONS; Cold Weather Brings Surge In Prices of Heating Fuels, Retrieved on 2008-05-25
81. BBC, Heatwave causes electricity surge, Retrieved on 2008-05-25
82. The Seven Key Messages of the Energy Drill Program, Retrieved on 2008-05-25
83. An international version called the Aeronautical Information Publication contains parallel information, as well as specific information on the international airports for use by the international community.

84. 7-1-22. PIREPs Relating to Airframe Icing, February 16, 2006, Aeronautical Information Manual, FAA AIM Online
85. Agricultural and Forest Meteorology, Elsevier, ISSN 0168-1923.
86. Encyclopædia Britannica, 2007.
87. About the HPC, NOAA/ National Weather Service, National Centers for Environmental Prediction, Hydrometeorological Prediction Center, Camp Springs, Maryland, 2007.
88. Super User. Home. Retrieved 16 June 2015.
89. DRIHM News, number 1, March 2012, p2 An ideal environment for hydrometeorology research at the European level
90. Modern research in nuclear meteorology (PDF). Atomic Energy. Springer New York. February 1974. doi:10.1007/BF01117823. Retrieved July 6, 2008.
91. Robert Hart and Jenni Evans (2003). Synoptic Composites of the Extratropical Transition Lifecycle of North Atlantic TCs as Defined Within Cyclone Phase Space (PDF). American Meteorological Society. Retrieved 2006-10-03. External link in |publisher= (help)
92. Ryan N. Maue. CHAPTER 3: CYCLONE PARADIGMS AND EXTRATROPICAL TRANSITION CONCEPTUALIZATIONS. Retrieved on 15 June 2008.
93. Atlantic Oceanographic and Meteorological Laboratory, Hurricane Research Division. Frequently Asked Questions: What is an extra-tropical cyclone?. NOAA. Retrieved 2006-07-25.
94. Office of the Federal Coordinator for Meteorology (2006). Appendix G: Glossary. NOAA. Retrieved on 2009-02-16.
95. Jack Williams (2007). What›s happening inside highs and lows. USA Today. Retrieved on 2009-02-16.
96. Myanmar government (2007). Haze. Retrieved on 2007-02-11.
97. Robert Tardif (2002). Fog characteristics. NCAR National Research Laboratory. Retrieved on 2007-02-11.
98. CBC News (2009). Blame Yukon: Arctic air mass chills rest of North America. Canadian Broadcasting Centre. Retrieved on 2009-02-16.
99. Glossary of Meteorology (2009). Level of nondivergence. American Meteorological Society. Retrieved on 2009-02-17.
100. Konstantin Matchev (2009). Middle-Latitude Cyclones - II. University of Florida. Retrieved on 2009-02-16.
101. Keith C. Heidorn (2005). Weather›s Highs and Lows: Part 1 The High. The Weather Doctor. Retrieved on 2009-02-16.
102. Instituto Nacional de Meteorologia. Meteorologia del Aeropuerto de la Palma. Retrieved on 2007-05-05.
103. Glossary of Meteorology (2009). High. American Meteorological Society. Retrieved on 2009-02-16.

104. Author unknown. Lesson 7: Clouds and Precipitation. Self-published. Archived from the original on January 11, 2005. Retrieved 2007-04-29.
105. David Roth. Unified Surface Analysis Manual (PDF). Hydrometeorological Prediction Center. Retrieved 2006-10-22.
106. Lynch, Peter (March 2008). The origins of computer weather prediction and climate modeling (PDF). Journal of Computational Physics. University of Miami. 227 (7): 3431–44. Bibcode:2008JCoPh.227.3431L. doi:10.1016/j.jcp.2007.02.034. Retrieved 2010-12-23.
107. Lynch, Peter (2006). Weather Prediction by Numerical Process. The Emergence of Numerical Weather Prediction. Cambridge University Press. pp. 1–27. ISBN 978-0-521-85729-1.
108. Charney, Jule; Fjørtoft, Ragnar; von Neumann, John (November 1950). Numerical Integration of the Barotropic Vorticity Equation. Tellus. 2 (4): 237. Bibcode:1950Tell....2..237C. doi:10.1111/j.2153-3490.1950.tb00336.x.
109. Cox, John D. (2002). Storm Watchers. John Wiley & Sons, Inc. p. 208. ISBN 0-471-38108-X.
110. Harper, Kristine; Uccellini, Louis W.; Kalnay, Eugenia; Carey, Kenneth; Morone, Lauren (May 2007). 2007: 50th Anniversary of Operational Numerical Weather Prediction. Bulletin of the American Meteorological Society. 88 (5): 639–650. Bibcode:2007BAMS...88..639H. doi:10.1175/BAMS-88-5-639.
111. American Institute of Physics (2008-03-25). Atmospheric General Circulation Modeling. Archived from the originalon 2008-03-25. Retrieved 2008-01-13.
112. Phillips, Norman A. (April 1956). The general circulation of the atmosphere: a numerical experiment. Quarterly Journal of the Royal Meteorological Society. 82 (352): 123–154. Bibcode:1956QJRMS..82..123P. doi:10.1002/qj.49708235202.
113. Cox, John D. (2002). Storm Watchers. John Wiley & Sons, Inc. p. 210. ISBN 0-471-38108-X.
114. Lynch, Peter (2006). The ENIAC Integrations. The Emergence of Numerical Weather Prediction. Cambridge University Press. pp. 206–208. ISBN 978-0-521-85729-1.
115. National Oceanic and Atmospheric Administration (2008-05-22). The First Climate Model. Retrieved 2011-01-08.
116. Leslie, L.M.; Dietachmeyer, G.S. (December 1992). Real-time limited area numerical weather prediction in Australia: a historical perspective (PDF). Australian Meteorological Magazine. Bureau of Meteorology. 41 (SP): 61–77. Retrieved 2011-01-03.
117. Shuman, Frederick G. (September 1989). History of Numerical Weather Prediction at the National Meteorological Center. Weather and Forecasting. 4 (3): 286–296. Bibcode:1989WtFor...4..286S. doi:10.1175/1520-0434(1989)004<0286:HONWPA>2.0.CO;2. ISSN 1520-0434.
118. Steyn, D. G. (1991). Air pollution modeling and its application VIII, Volume 8. Birkhäuser. pp. 241–242. ISBN 978-0-306-43828-8.

119. Xue, Yongkang; Fennessey, Michael J. (1996-03-20). Impact of vegetation properties on U. S. summer weather prediction (PDF). Journal of Geophysical Research. American Geophysical Union. 101 (D3): 7419. Bibcode:1996JGR...101.7419X. doi:10.1029/95JD02169. Retrieved 2011-01-06.

120. Hughes, Harry (1976). Model output statistics forecast guidance. United States Air Force Environmental Technical Applications Center. pp. 1–16.

121. Best, D. L.; Pryor, S. P. (1983). Air Weather Service Model Output Statistics Systems. Air Force Global Weather Central. pp. 1–90.

122. Toth, Zoltan; Kalnay, Eugenia (December 1997). Ensemble Forecasting at NCEP and the Breeding Method. Monthly Weather Review. 125 (12): 3297–3319. Bibcode:1997MWRv..125.3297T. doi:10.1175/1520-0493(1997)125<3297:EFANAT>2.0.CO;2. ISSN 1520-0493.

123. The Ensemble Prediction System (EPS). ECMWF. Archived from the original on 2010-10-30. Retrieved 2011-01-05.

124. Molteni, F.; Buizza, R.; Palmer, T.N.; Petroliagis, T. (January 1996). The ECMWF Ensemble Prediction System: Methodology and validation. Quarterly Journal of the Royal Meteorological Society. 122 (529): 73–119. Bibcode:1996QJRMS.122...73M. doi:10.1002/qj.49712252905.

125. Stensrud, David J. (2007). Parameterization schemes: keys to understanding numerical weather prediction models. Cambridge University Press. p. 56. ISBN 978-0-521-86540-1. Retrieved 2011-02-15.

126. National Climatic Data Center (2008-08-20). Key to METAR Surface Weather Observations. National Oceanic and Atmospheric Administration. Retrieved 2011-02-11.

127. SYNOP Data Format (FM-12): Surface Synoptic Observations. UNISYS. 2008-05-25. Archived from the original on 2007-12-30.

128. Krishnamurti, T. N. (January 1995). Numerical Weather Prediction. Annual Review of Fluid Mechanics. 27 (1): 195–225. Bibcode:1995AnRFM..27..195K. doi:10.1146/annurev.fl.27.010195.001211.

129. The WRF Variational Data Assimilation System (WRF-Var). University Corporation for Atmospheric Research. 2007-08-14. Archived from the original on 2007-08-14.

130. Gaffen, Dian J. (2007-06-07). Radiosonde Observations and Their Use in SPARC-Related Investigations. Archived from the original on 2007-06-07.

131. Ballish, Bradley A.; V. Krishna Kumar (November 2008). Systematic Differences in Aircraft and Radiosonde Temperatures (PDF). Bulletin of the American Meteorological Society. 89 (11): 1689–1708. Bibcode:2008BAMS...89.1689B. doi:10.1175/2008BAMS2332.1. Retrieved 2011-02-16.

132. National Data Buoy Center (2009-01-28). The WMO Voluntary Observing Ships (VOS) Scheme. National Oceanic and Atmospheric Administration. Retrieved 2011-02-15.

133. 403rd Wing (2011). The Hurricane Hunters. 53rd Weather Reconnaissance

Squadron. Retrieved 2006-03-30.

134. Lee, Christopher (2007-10-08). Drone, Sensors May Open Path Into Eye of Storm. The Washington Post. Retrieved 2008-02-22.

135. National Oceanic and Atmospheric Administration (2010-11-12). NOAA Dispatches High-Tech Research Plane to Improve Winter Storm Forecasts. Retrieved 2010-12-22.

136. Stensrud, David J. (2007). Parameterization schemes: keys to understanding numerical weather prediction models. Cambridge University Press. p. 137. ISBN 978-0-521-86540-1. Retrieved 2011-01-08.

137. Houghton, John Theodore (1985). The Global Climate. Cambridge University Press archive. pp. 49–50. ISBN 978-0-521-31256-1. Retrieved 2011-01-08.

138. Pielke, Roger A. (2002). Mesoscale Meteorological Modeling. Academic Press. pp. 48–49. ISBN 0-12-554766-8.

139. Pielke, Roger A. (2002). Mesoscale Meteorological Modeling. Academic Press. pp. 18–19. ISBN 0-12-554766-8.

140. Strikwerda, John C. (2004). Finite difference schemes and partial differential equations. SIAM. pp. 165–170. ISBN 978-0-89871-567-5. Retrieved 2010-12-31.

141. Pielke, Roger A. (2002). Mesoscale Meteorological Modeling. Academic Press. p. 65. ISBN 0-12-554766-8.

142. Pielke, Roger A. (2002). Mesoscale Meteorological Modeling. Academic Press. pp. 285–287. ISBN 0-12-554766-8.

143. Sunderam, V. S.; van Albada, G. Dick; Peter, M. A.; Sloot, J. J. Dongarra (2005). Computational Science – ICCS 2005: 5th International Conference, Atlanta, GA, USA, May 22–25, 2005, Proceedings, Part 1. Springer. p. 132. ISBN 978-3-540-26032-5. Retrieved 2011-01-02.

144. Zwieflhofer, Walter; Kreitz, Norbert; European Centre for Medium Range Weather Forecasts (2001). Developments in teracomputing: proceedings of the ninth ECMWF Workshop on the Use of High Performance Computing in Meteorology. World Scientific. p. 276. ISBN 978-981-02-4761-4. Retrieved 2011-01-02.

145. Chan, Johnny C. L. & Jeffrey D. Kepert (2010). Global Perspectives on Tropical Cyclones: From Science to Mitigation. World Scientific. pp. 295–296. ISBN 978-981-4293-47-1. Retrieved 2011-02-24.

146. Holton, James R. (2004). An introduction to dynamic meteorology, Volume 1. Academic Press. p. 480. ISBN 978-0-12-354015-7. Retrieved 2011-02-24.

147. Brown, Molly E. (2008). Famine early warning systems and remote sensing data. Springer. p. 121. ISBN 978-3-540-75367-4. Retrieved 2011-02-24.

148. Ahrens, C. Donald (2008). Essentials of meteorology: an invitation to the atmosphere. Cengage Learning. p. 244. ISBN 978-0-495-11558-8. Retrieved 2011-02-11.

149. Narita, Masami & Shiro Ohmori (2007-08-06). 3.7 Improving Precipitation Forecasts by the Operational Nonhydrostatic Mesoscale Model with the Kain-Fritsch Convective Parameterization and Cloud Microphysics(PDF). 12th

Conference on Mesoscale Processes. American Meteorological Society. Retrieved 2011-02-15.

150. Frierson, Dargan (2000-09-14). The Diagnostic Cloud Parameterization Scheme (PDF). University of Washington. pp. 4–5. Retrieved 2011-02-15.
151. Stensrud, David J. (2007). Parameterization schemes: keys to understanding numerical weather prediction models. Cambridge University Press. p. 6. ISBN 978-0-521-86540-1. Retrieved 2011-02-15.
152. McGuffie, K. & A. Henderson-Sellers (2005). A climate modelling primer. John Wiley and Sons. p. 188. ISBN 978-0-470-85751-9.
153. Melnikova, Irina N. & Alexander V. Vasilyev (2005). Short-wave solar radiation in the earth›s atmosphere: calculation, oberservation, interpretation. Springer. pp. 226–228. ISBN 978-3-540-21452-6.
154. Stensrud, David J. (2007). Parameterization schemes: keys to understanding numerical weather prediction models. Cambridge University Press. pp. 12–14. ISBN 978-0-521-86540-1. Retrieved 2011-02-15.
155. Baklanov, Alexander, Sue Grimmond, Alexander Mahura (2009). Meteorological and Air Quality Models for Urban Areas. Springer. pp. 11–12. ISBN 978-3-642-00297-7. Retrieved 2011-02-24.
156. Warner, Thomas Tomkins (2010). Numerical Weather and Climate Prediction. Cambridge University Press. p. 259. ISBN 978-0-521-51389-0. Retrieved 2011-02-11.
157. Lynch, Peter (2006). The Fundamental Equations. The Emergence of Numerical Weather Prediction. Cambridge University Press. pp. 45–46. ISBN 978-0-521-85729-1.
158. Ahrens, C. Donald (2008). Essentials of meteorology: an invitation to the atmosphere. Cengage Learning. p. 10. ISBN 978-0-495-11558-8. Retrieved 2011-02-11.
159. Janjic, Zavisa; Gall, Robert; Pyle, Matthew E. (February 2010). Scientific Documentation for the NMM Solver(PDF). National Center for Atmospheric Research. pp. 12–13. Retrieved 2011-01-03.
160. Pielke, Roger A. (2002). Mesoscale Meteorological Modeling. Academic Press. pp. 131–132. ISBN 0-12-554766-8.
161. Baum, Marsha L. (2007). When nature strikes: weather disasters and the law. Greenwood Publishing Group. p. 189. ISBN 978-0-275-22129-4. Retrieved 2011-02-11.
162. Glahn, Harry R.; Lowry, Dale A. (December 1972). The Use of Model Output Statistics (MOS) in Objective Weather Forecasting. Journal of Applied Meteorology. 11 (8): 1203–1211. Bibcode:1972JApMe..11.1203G. doi:10.1175/1520-0450(1972)011<1203:TUOMOS>2.0.CO;2. ISSN 1520-0450.
163. Gultepe, Ismail (2007). Fog and boundary layer clouds: fog visibility and forecasting. Springer. p. 1144. ISBN 978-3-7643-8418-0. Retrieved 2011-02-11.
164. Barry, Roger Graham; Chorley, Richard J. (2003). Atmosphere, weather, and

climate. Psychology Press. p. 172. ISBN 978-0-415-27171-4. Retrieved 2011-02-11.

165. Cox, John D. (2002). Storm Watchers. John Wiley & Sons, Inc. pp. 222–224. ISBN 0-471-38108-X.

166. Manousos, Peter (2006-07-19). Ensemble Prediction Systems. Hydrometeorological Prediction Center. Retrieved 2010-12-31.

167. Weickmann, Klaus; Jeff Whitaker; Andres Roubicek; Catherine Smith (2001-12-01). The Use of Ensemble Forecasts to Produce Improved Medium Range (3–15 days) Weather Forecasts.. Climate Diagnostics Center. Retrieved 2007-02-16.

168. Chakraborty, Arindam (October 2010). The Skill of ECMWF Medium-Range Forecasts during the Year of Tropical Convection 2008. Monthly Weather Review. 138 (10): 3787–3805. Bibcode:2010MWRv..138.3787C. doi:10.1175/2010MWR3217.1.

169. Epstein, E.S. (December 1969). Stochastic dynamic prediction. Tellus A. 21 (6): 739–759. Bibcode:1969Tell...21..739E. doi:10.1111/j.2153-3490.1969.tb00483.x.

170. Leith, C.E. (June 1974). Theoretical Skill of Monte Carlo Forecasts. Monthly Weather Review. 102 (6): 409–418. Bibcode:1974MWRv..102..409L. doi:10.1175/1520-0493(1974)102<0409:TSOMCF>2.0.CO;2. ISSN 1520-0493.

171. MOGREPS. Met Office. Retrieved 2012-11-01.

172. Warner, Thomas Tomkins (2010). Numerical Weather and Climate Prediction. Cambridge University Press. pp. 266–275. ISBN 978-0-521-51389-0. Retrieved 2011-02-11.

173. Palmer, T.N.; Shutts, G.J.; Hagedorn, R.; Doblas-Reyes, F.J.; Jung, T.; Leutbecher, M. (May 2005). Representing Model Uncertainty in Weather and Climate Prediction. Annual Review of Earth and Planetary Sciences. 33: 163–193. Bibcode:2005AREPS..33..163P. doi:10.1146/annurev.earth.33.092203.122552. Retrieved 2011-02-09.

174. Grimit, Eric P.; Mass, Clifford F. (October 2004). Redefining the Ensemble Spread-Skill Relationship from a Probabilistic Perspective (PDF). University of Washington. Retrieved 2010-01-02.

175. Zhou, Binbin; Du, Jun (February 2010). Fog Prediction From a Multimodel Mesoscale Ensemble Prediction System (PDF). Weather and Forecasting. American Meteorological Society. 25: 303. Bibcode:2010WtFor..25..303Z. doi:10.1175/2009WAF2222289.1. Retrieved 2011-01-02.

176. Cane, D.; Milelli, M. (2010-02-12). Multimodel SuperEnsemble technique for quantitative precipitation forecasts in Piemonte region (PDF). Natural Hazards and Earth System Sciences. 10 (2): 265. Bibcode:2010NHESS..10..265C. doi:10.5194/nhess-10-265-2010. Retrieved 2011-01-02.

177. Daly, Aaron & Paolo Zannetti (2007). Ambient Air Pollution (PDF). The Arab School for Science and Technology and The EnviroComp Institute. p. 16. Retrieved 2011-02-24.

178. Baklanov, Alexander; Rasmussen, Alix; Fay, Barbara; Berge, Erik; Finardi, Sandro

(September 2002). Potential and Shortcomings of Numerical Weather Prediction Models in Providing Meteorological Data for Urban Air Pollution Forecasting. Water, Air, & Soil Pollution: Focus. 2 (5): 43–60. doi:10.1023/A:1021394126149.

179. Marshall, John; Plumb, R. Alan (2008). Atmosphere, ocean, and climate dynamics: an introductory text. Amsterdam: Elsevier Academic Press. pp. 44–46. ISBN 978-0-12-558691-7.
180. Australian Bureau of Statistics (2005). Year book, Australia, Issue 87. p. 40. Retrieved 2011-02-18.
181. National Oceanic and Atmospheric Administration 200th Celebration (2008-05-22). The First Climate Model. National Oceanic and Atmospheric Administration. Retrieved 2010-04-20.
182. Bridgman, Howard A., John E. Oliver, Michael H. Glantz (2006). The global climate system: patterns, processes, and teleconnections. Cambridge University Press. pp. 284–289. ISBN 978-0-521-82642-6. Retrieved 2011-02-18.
183. Chalikov, D. V. (August 1978). The numerical simulation of wind-wave interaction. Journal of Fluid Mechanics. 87 (3): 561–82. Bibcode:1978JFM....87..561C. doi:10.1017/S0022112078001767.
184. Lin, Pengzhi (2008). Numerical modeling of water waves. Psychology Press. p. 270. ISBN 978-0-415-41578-1.
185. Bender, Leslie C. (January 1996). Modification of the Physics and Numerics in a Third-Generation Ocean Wave Model. Journal of Atmospheric and Oceanic Technology. 13 (3): 726. Bibcode:1996JAtOT..13..726B. doi:10.1175/1520-0426(1996)013<0726:MOTPAN>2.0.CO;2. ISSN 1520-0426.
186. National Hurricane Center (July 2009). Technical Summary of the National Hurricane Center Track and Intensity Models (PDF). National Oceanic and Atmospheric Administration. Retrieved 2011-02-19.
187. Franklin, James (2010-04-20). National Hurricane Center Forecast Verification. National Hurricane Center. Retrieved 2011-01-02.
188. Rappaport, Edward N.; Franklin, James L.; Avila, Lixion A.; Baig, Stephen R.; Beven II, John L.; Blake, Eric S.; Burr, Christopher A.; Jiing, Jiann-Gwo; Juckins, Christopher A.; Knabb, Richard D.; Landsea, Christopher W.; Mainelli, Michelle; Mayfield, Max; McAdie, Colin J.; Pasch, Richard J.; Sisko, Christopher; Stewart, Stacy R.; Tribble, Ahsha N. (April 2009). Advances and Challenges at the National Hurricane Center. Weather and Forecasting. 24 (2): 395–419. Bibcode:2009WtFor..24..395R. doi:10.1175/2008WAF2222128.1.
189. Sullivan, Andrew L. (June 2009). Wildland surface fire spread modelling, 1990–2007. 1: Physical and quasi-physical models. International Journal of Wildland Fire. 18 (4): 349. arXiv:0706.3074 . doi:10.1071/WF06143.
190. Asensio, M. I. & L. Ferragut (2002). On a wildland fire model with radiation. International Journal for Numerical Methods in Engineering. 54: 137–157. Bibcode:2002IJNME..54..137A. doi:10.1002/nme.420.
191. Mandel, Jan, Lynn S. Bennethum, Jonathan D. Beezley, Janice L. Coen, Craig

C. Douglas, Minjeong Kim, and Anthony Vodacek (2008). A wildfire model with data assimilation. Mathematics and Computers in Simulation. 79 (3): 584–606. arXiv:0709.0086 . Bibcode:2007arXiv0709.0086M. doi:10.1016/j.matcom.2008.03.015.

192. Clark, T. L., M. A. Jenkins, J. Coen, and David Packham (1996). A coupled atmospheric-fire model: Convective Froude number and dynamic fingering. International Journal of Wildland Fire. 6 (4): 177–190. doi:10.1071/WF9960177.

193. Clark, Terry L., Marry Ann Jenkins, Janice Coen, and David Packham (1996). A coupled atmospheric-fire model: Convective feedback on fire line dynamics. Journal of Applied Meteorology. 35 (6): 875–901. Bibcode:1996JApMe..35..875C. doi:10.1175/1520-0450(1996)035<0875:ACAMCF>2.0.CO;2. ISSN 1520-0450.

194. Rothermel, Richard C. (January 1972). A mathematical model for predicting fire spread in wildland fires (PDF). United States Forest Service. Retrieved 2011-02-28.

CHAPTER 6

Atmospheric Circulation

Chapter Contents

Atmospheric circulation is the large-scale movement of air by which heat is distributed on the surface of the Earth. The wind belts and the jet streams girdling the planet are steered by three convection cells: the Hadley cell, the Ferrel cell, and the Polar cell.

INTRODUCTION

Atmospheric circulation is the large-scale movement of air, and together with ocean circulation is the means by which thermal energy is redistributed on the surface of the Earth.

The Earth's atmospheric circulation varies from year to year, but the large scale structure of its circulation remains fairly constant. The smaller scale weather systems – mid-latitude depressions, or tropical convective cells – occur "randomly", and long range weather predictions of those cannot be made beyond ten days in practice, or a month in theory. The large scale atmospheric circulation of the Earth, however, is an average of its systems and patterns, and is considered stable over longer periods of time.

The Earth's weather is a consequence of its illumination by the Sun, and the laws of thermodynamics. The atmospheric circulation can be viewed, from that standpoint, as a heat engine driven by the Sun's energy, and whose energy sink, ultimately, is the blackness of space. The work produced by that engine causes the motion of the masses of air and in that process it redistributes the energy absorbed by the Earth's surface near the tropics to space and incidentally to the latitudes nearer the poles.

The large scale atmospheric circulation "cells" shift polewards in warmer periods (e.g. interglacials compared to glacials), but remain largely constant as they are, fundamentally, a property of the Earth's size, rotation rate, heating and atmospheric depth, all of which change little. Over very long time periods (hundreds of millions of years), a tectonic uplift can significantly alter their major elements, such as the jet stream, and plate tectonics may shift ocean currents. During the extremely hot climates of the Mesozoic, a third desert belt may have existed at the Equator. But, the overall latitudinal pattern of Earth's climate has not changed.

Latitudinal Circulation Features

The wind belts girdling the planet are organised into three cells; they are from the equator toward the poles, the Hadley cell, the Ferrel cell, and the Polar cell. Those cells exist in both the northern and southern hemispheres. The vast bulk of the atmospheric motion occurs in the Hadley cell. Note that there is one discrete Hadley cell that may split, shift and merge in a complicated process over time. The high pressure systems acting on the Earth's surface is balanced by the low pressure systems elsewhere. As a result, there a balance of forces acting on the Earth's surface.

HADLEY CELL

The Hadley cell, named after George Hadley, is a tropical atmospheric circulation that features air rising near the equator, flowing poleward at 10–15 kilometers above the surface, descending in the subtropics, and then flowing equatorward near the surface. This circulation creates the trade winds, tropical rain-belts and hurricanes, subtropical deserts and the jet streams.

There is one primary circulation cell known as a Hadley cell and two secondary circulation cells at higher latitudes, known as the Ferrel cell, and Polar cell. It can consist from 30 to 40 degrees north and south and is mainly responsible for the weather in the equatorial regions of the world.

Mechanism

The major driving force of atmospheric circulation is solar heating, which is greatest near the equator and lesser at the poles. The atmospheric circulation transports energy polewards, thus reducing the resulting equator-to-pole temperature gradient. The mechanisms by which this is accomplished differs in tropical and extratropical latitudes.

Hadley cells exist on either side of the equator. Each cell encircles the globe and acts to transport energy from the equator to 30th latitude. This is the result of the buoyancy of warm air rising from the equator to the tropopause; its movement as it is pushed poleward by air rising from below; its cooling as it moves poleward at high altitude; its sinking near subtropics (30th parallels); and finally its equatorward return near the surface, completing the loop. In higher latitudes, the energy transport is also accomplished by cyclones and anticyclones that cause relatively warm air to move polewards and cold air to move equatorwards. The large scale tropical overturning cell is referred to as the Hadley cell. Why it extends only to 30 degrees latitude and what determines its strength are questions addressed by dynamical meteorology.

Near the tropopause, as the air moves polewards in the Hadley cell, it is turned eastward by the Coriolis effect, creating the subtropical jet streams that flow from west to east. Analogously, near the surface, the equatorward return flow is turned to the west by the Coriolis effect. These resulting surface winds, with both an equatorward and a westward component, are referred to as the trade winds.

The Hadley system provides an example of a thermally direct circulation. The thermodynamic efficiency of the Hadley system, considered as a heat engine, has been relatively constant over the 1979~2010 period, averaging 2.6%. Over the same interval, the power generated by the Hadley regime has risen at an average rate of about 0.54 TW per yr; this reflects an increase in energy input to the system consistent with the observed trend in the tropical sea surface temperatures.

History of Discovery

In the early 18th century, George Hadley, an English lawyer and amateur meteorologist, was dissatisfied with the theory that the astronomer Edmond Halley had proposed for explaining the trade winds. What was no doubt correct in Halley's theory was that solar heating creates upward motion of equatorial air, and air mass from neighboring latitudes must flow in to replace the risen air mass. But for the westward component of the trade winds Halley had proposed that in moving across the sky the Sun heats the air mass differently over the course of the day. Hadley was not satisfied with that part of Halley's theory and rightly so. Hadley was the first to recognize that Earth's rotation plays a role in the direction taken by an air mass as it moves relative to the Earth. Hadley's theory, published in 1735, remained unknown, but it was rediscovered independently several times. Among the re-discoverers was John Dalton, who later learned of Hadley's priority. Over time the mechanism proposed by Hadley became accepted, and over time his name was increasingly attached to it. By the end of the 19th century it was shown that Hadley's theory was deficient in several respects. One of the first who accounted for the dynamics correctly was William Ferrel. It took many decades for the correct theory to become accepted, and even today Hadley's theory can still be encountered occasionally, particularly in popular books and websites. Hadley's theory was the generally accepted theory long enough to make his name become universally attached to the circulation pattern in the tropical atmosphere. In 1980 Isaac Held and Arthur Hou developed the Held-Hou Model to describe the Hadley circulation.

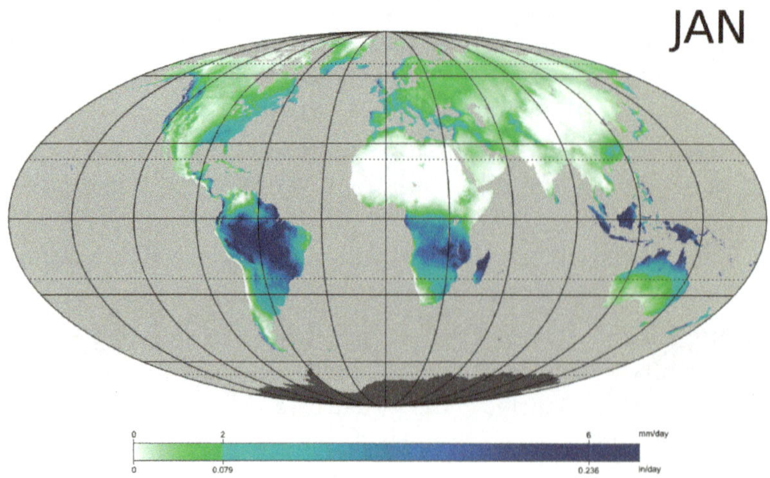

Long-term mean precipitation by month

Cloud formations in a famous image of Earth from Apollo 17, makes similar circulation directly visible

Major Impacts on Precipitation by Latitude

The region in which the equatorward moving air masses converge and rise, is known as the intertropical convergence zone, or ITCZ. Within that zone develops a band of thunderstorms that produce high-precipitation.

Having lost most of its water vapor to condensation and precipitation in the upward branch of the Hadley cell circulation, the descending air is dry. As the air descends, low relative humidities are produced as the air is warmed adiabatically by compression from the overlying air, producing a region of higher pressure. The subtropics are relatively free of the convection, or thunderstorms, that are common in the equatorial belt. Many of the world's deserts are located in these subtropical latitudes.

Hadley Cell Expansion

There is some evidence that the expansion of the Hadley cells is related to climate change. The majority of Earth's arid regions are located in the areas beneath the descending part of the Hadley circulation at around 30 degrees latitude. Those models show that the Hadley cell will expand with increased global mean temperature (perhaps by 2 degrees latitude over the 21st century). This might lead to large changes in precipitation in the latitudes at the edge of the cells. Scientists fear that global warming might bring changes to the ecosystems in the deep tropics and that the deserts will become drier and expand. As the areas around 30 degrees latitude become drier, those inhabiting that region will see less rainfall than traditionally expected, which could cause difficulty with food supplies and livability. There is strong evidence of paleoclimate climate change in central Africa's rain forest in c. 850 B.C. Palynological (fossil pollen) evidence shows a drastic change in rain forest biome to that of open savannah as a consequence of wide-scale drying not connected necessarily to intermittent drought but perhaps to gradual warming. The hypothesis, that a decline in solar activity reduces the latitudinal extent of the Hadley Circulation and decreases mid-latitudinal monsoon intensity, is matched by data, showing increased dryness in central west Africa and increase in precipitation in temperate zones north. Meanwhile, mid-latitudinal storm tracks in the temperate zones increased and moved equatorward.

Polar Cell

The **Polar cell**, likewise, is a simple system. Though cool and dry relative to equatorial air, the air masses at the 60th parallel are still sufficiently warm and moist to undergo convection and drive a thermal loop. At the 60th parallel, the air rises to the tropopause (about 8 km at this latitude) and moves poleward. As it does so, the upper level air mass deviates toward the east. When the air reaches the polar areas, it has cooled and is considerably denser than the underlying air. It descends, creating a cold, dry high-pressure area. At the polar surface level, the mass of air is driven toward the 60th parallel, replacing the air that rose there, and the Polar circulation cell is complete. As the air at the surface moves equatorward, it deviates toward the west. Again, the deviations of the air masses is due to conservation of energy, which is also referred to as the Coriolis effect. The air flows at the surface are called the Polar easterlies (easterly from the east).

The outflow of air mass from the cell creates harmonic waves in the atmosphere known as Rossby waves. These ultra-long waves determine the path of the jet stream, which travels within the transitional zone between the tropopause and the Ferrel cell. By acting as a heat sink, the Polar cell moves the abundant heat from the equator toward the polar regions.

The Hadley cell and the Polar cell are similar in that they are thermally direct; in other words, they exist as a direct consequence of surface temperatures. Their thermal characteristics drive the weather in their domain. The sheer volume of energy that the Hadley cell transports, and the depth of the heat sink that is the Polar cell, ensures that the effects of transient weather phenomena are not only not felt by the system as a whole, but — except under unusual circumstances — do not form. The endless chain of passing highs and lows which is part of everyday life for mid-latitude dwellers, at latitudes between 30 and 60° latitude, is unknown above the 60th and below the 30th parallels. There are some notable exceptions to this rule. In Europe, unstable weather extends to at least 70° north.

These atmospheric features are stable. Even though they may strengthen or weaken regionally over time, they do not vanish entirely.

The Polar cell, orography and Katabatic winds in Antarctica, can create very cold conditions at the surface, for instance the coldest temperature recorded on Earth: −89.2 °C at Vostok Station in Antarctica, measured 1983.

Ferrel Cell

Part of the air rising at 60° latitude diverges at high altitude toward the poles and creates the polar cell. The rest moves toward the equator where it collides at 30° latitude with the high level air of the Hadley cell. There it subsides and strengthens the high pressure ridges beneath. A large part of the energy that drives the Ferrel cell is provided by the Polar and Hadley cells circulating on either side and that drag the Ferrel cell with it. The Ferrel cell, theorized by William Ferrel (1817–1891), is therefore a secondary circulation feature, whose existence depends upon the Hadley and Polar cells on either side of it; it behaves much as an atmospheric ball bearing between the two. It might be thought of as an eddy created by the Hadley and Polar cells. The Ferrel cell is weak and the air flow and temperatures within are variable. For this reason, the mid latitudes are sometimes known as the **"zone of mixing."** At high altitudes, the Ferrel cell overrides the Hadley and Polar cells. The air of the Ferrel cell that descends at 30° latitude returns poleward at the ground level and as it does so it deviates toward the east. In the upper atmosphere of the Ferrel cell the air moving equatorward, deviates toward the west. Both of those deviations, as in the case of the Hadley and Polar cells, are driven by conservation of energy. As a result, just as the Easterly Trade Winds are found below the Hadley cell, the Westerlies are found beneath the Ferrel cell. The forces driving the flow in the Ferrel cell are weak and so the weather in that zone is variable. Thus, strong high-pressure areas which divert the prevailing westerlies, such as a Siberian high, can override the Ferrel cell, making it discontinuous.

While the Hadley and Polar cells are truly closed loops, the Ferrel cell is not, and the telling point is in the Westerlies, which are more formally known as «the Prevailing Westerlies.» The Easterly Trade Winds and the Polar Easterlies have nothing over which to prevail, as their parent circulation cells are strong enough, and face few obstacles either in the form of massive terrain features or high pressure zones. The weaker Westerlies of the Ferrel cell, however, can be disrupted. The local passage of a cold front may change that in a matter of minutes, and frequently does. As a result, at the surface, winds can vary abruptly in direction. But the winds above the surface, where they are less disrupted by terrain, are essentially westerly. A low pressure zone at 60° latitude that moves equator wards, or a high pressure zone at 30° latitude that moves poleward, will accelerate the Westerlies of the Ferrel cell. A strong high, moving polewards may bring Westerly winds for days. As the high and low pressure zones are established by the Hadley and Polar cells, it can be said that they drive the weather of the middle latitude Ferrel cell.

The Ferrel cell is driven by the Hadley and Polar cells. It has neither a strong source of heat nor a strong cold sink to drive convection. As a result, the weather within the Ferrel cell is highly variable and is influenced by changes to the Hadley and Polar cells. The base of the Ferrel cell is characterized by the movement of air masses, and the location of those air masses is influenced in part by the location of the jet stream, even though it flows near the tropopause. Overall, the movement of surface air is from the 30th latitude to the 60th. However, the upper flow of the Ferrel cell is weak and not well defined.

In contrast to the Hadley and Polar systems, the Ferrel system provides an example of a thermally indirect circulation. The Ferrel system acts as a heat pump with a coefficient of performance of 12.1, consuming kinetic energy at an approximate rate of 275 TW.

Longitudinal Circulation Features

While the Hadley, Ferrel, and Polar cells (whose axes are oriented along parallels i.e. latitudes) are the major features of global heat transport, they do not act alone. Temperature differences also drive a set of circulation cells, whose axis of circulation are longitudinally oriented. This atmospheric motion is known as zonal overturning circulation.

Latitudinal circulation is a result of the highest solar radiation per unit area (solar intensity) falling on the tropics. The solar intensity decreases as the latitude increases, reaching essentially zero at the poles. Longitudinal circulation, however, is a result of the heat capacity of water, it's absorptivity, and it's mixing. Water absorbs more heat than does the land, but its temperature does not rise as greatly as does the land. As a result, temperature variations on land are greater than on water. The Hadley, Ferrel, and Polar cells operate at the largest scale of thousands of kilometers (synoptic scale). But, even at mesoscales (a horizontal range of 5 to several hundred kilometres), this effect is noticeable. During the day, air warmed by the relatively hotter land rises, and as it does so it draws a cool breeze from the sea that replaces the risen air. At night, the relatively warmer water and cooler land reverses the process, and a breeze from the land, of air cooled by the land, is carried offshore by night. This described effect is daily (diurnal).

At the larger, synoptic, scale of oceans and continents, this effect is seasonal or even decadal. Warm air rising over the equatorial, continental, and western Pacific Ocean regions. When it reaches the tropopause, it cools and subsides in a region of relatively cooler water mass.

The Pacific Ocean cell plays a particularly important role in Earth's weather. This entirely ocean-based cell comes about as the result of a marked difference in the surface temperatures of the western and eastern Pacific. Under ordinary circumstances, the western Pacific waters are warm and the eastern waters are cool. The process begins when strong convective activity over equatorial East Asia and subsiding cool air off South America's west coast creates a wind pattern which pushes Pacific water westward and piles it up in the western Pacific. (Water levels in the western Pacific are about 60 cm higher than in the eastern Pacific, a difference due entirely to the force of moving air.)

WALKER CIRCULATION

The Walker circulation, also known as the Walker cell, is a conceptual model of the air flow in the tropics in the lower atmosphere (troposphere). According to this model, parcels of air follow a closed circulation in the zonal and vertical directions. This circulation, which is roughly consistent with observations, is caused by differences in heat distribution between ocean and land. It was discovered by Gilbert Walker. In addition to motions in the zonal and vertical direction the tropical atmosphere also has considerable motion in the meridional direction as part of, for example, the Hadley Circulation.

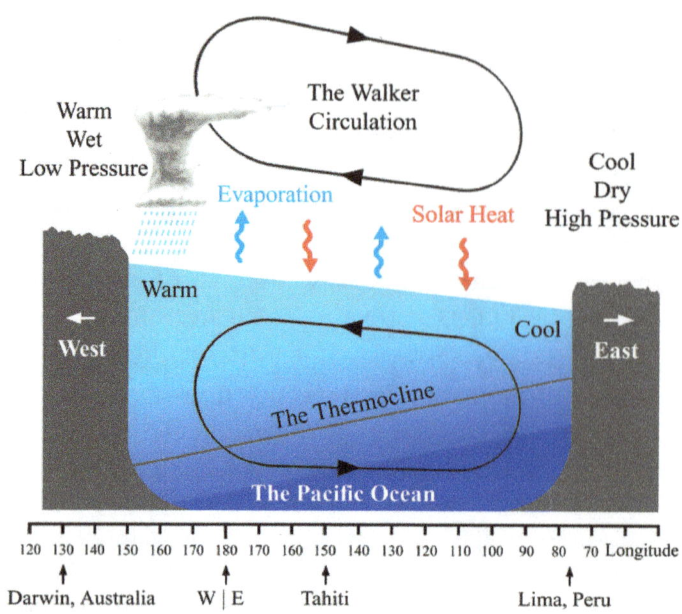

A schematic diagram of the quasi-equilibrium and La Niña phase of the southern oscillation. The Walker circulation is seen at the surface as easterly trade winds which move water and air warmed by the sun towards the west. The western side of the equatorial Pacific is characterized by warm, wet low pressure weather as the collected moisture is dumped in the form of typhoons and thunderstorms. The ocean is some 60 cm higher in the western Pacific as the result of this motion. The water and air are returned to the east. Both are now much cooler, and the air is much drier. An El Niño episode is characterised by a breakdown of this water and air cycle, resulting in relatively warm water and moist air in the eastern Pacific.

Walker's Methodology

Gilbert Walker was an established applied mathematician at the University of Cambridge when he became director-general of observatories in India in 1904. While there, he studied the characteristics of the Indian Ocean monsoon, the failure of whose rains had brought severe famine to the country in 1899. Analyzing vast amounts of weather data from India and the rest of the world, over the next fifteen years he published the first descriptions of the great seesaw oscillation of atmospheric pressure between the Indian and Pacific Ocean, and its correlation to temperature and rainfall patterns across much of the Earth's tropical regions, including India. He also worked with the Indian Meteorological Department especially in linking the monsoon with Southern Oscillation phenomenon. He was made a Companion of the Order of the Star of India in 1911.

Walker determined that the time scale of a year (used by many studying the atmosphere) was unsuitable because geospatial relationships could be entirely different depending on the season. Thus, Walker broke his temporal analysis into December–February, March–May, June–August, and September–November.

Walker then selected a number of "centers of action", which included areas such as the Indian Peninsula. The centers were in the hearts of regions with either permanent or seasonal high and low pressures. He also added points for regions where rainfall, wind or temperature was an important control.

He examined the relationships of the summer and winter values of pressure and rainfall, first focusing on summer and winter values, and later extending his work to the spring and autumn.

He concludes that variations in temperature are generally governed by variations in pressure and rainfall. It had previously been suggested that sunspots could be the cause of the temperature variations, but Walker argued against this conclusion by showing monthly correlations of sunspots with temperature, winds, cloud cover, and rain that were inconsistent.

Walker made it a point to publish all of his correlation findings, both of relationships found to be important as well as relationships that were found to be unimportant. He did this for the purpose of dissuading researchers from focusing on correlations that did not exist.

Mathematical Basis

The statistical model involved in the analysis of atmospheric data that led to the discovery of the Walker circulation is called an autoregressive (AR) process.

Autocorrelation Function

As background, first consider the autocorrelation function. An autocorrelation function is a measure of the dependence of time series values at one time on the values at another time. Given the time series x(n),n=1,2,...N, the autocorrelation function at lag k is defined as:

$$R_{xx}(k) = \frac{1}{(N-k)} \sum_{i=1}^{N-k} x(i)x(i+k).$$

The value of the autocorrelation function at lag 0 is the power of x(n), or its variance if the mean value of x(n) is zero:

$$R_{xx}(0) = \frac{1}{N} \sum_{i=1}^{N} x(i)^2.$$

Moreover, $\sqrt{R_{xx}(\infty)}$, is the mean value for random processes.

The autocorrelation function may be used to detect deterministic components masked in a random background because autocorrelation functions of deterministic data (like sine wave) persist over all time displacements, while autocorrelation functions of stochastic processes tend to zero for large time displacement (for 0-mean time series).

Autoregressive Model

Next, consider the autoregressive model proposed by Walker. Autoregressive Modeling is mathematical modeling of a time series based on the assumption that each value of the series depends only on a weighted sum of the previous values of the same series plus "noise".

If x(j) is the j-th value of the time series, the AR model of order p is given by:

$$x(j) = \sum_{i=1}^{p} a_i x(j-i) + n(j).$$

where n(j) is the noise. The order, p, can be considered as an index of the lag within the time series of which data will be considered for the analysis. The larger the lag, the larger the system of equations to be solved.

The AR coefficients can be estimated from the autocorrelation sequence by solving the Yule-Walker equations.

The generalized matrix version of the AR(p) model is given by the equation

$$X_t = \sum_{i=1}^{p} \varphi_i X_{t-i} + \varepsilon_t.$$

Gidon Eshel provides a useful breakdown of the Yule-Walker equations that discusses their relation to between the least squares approach for fitting an AR(p) model.

Yule-Walker Equations

The AR(p) model is given by the equation

$$X_t = \sum_{i=1}^{p} \varphi_i X_{t-i} + \varepsilon_t.$$

It is based on parameters φ_i where i=1,...,p. There is a direct correspondence between these parameters and the covariance function of the process, and this correspondence can be inverted to determine the parameters from the autocorrelation function (which is itself obtained from the covariances). This is done using the Yule-Walker equations:

$$\gamma_m = \sum_{i=1}^{p} \varphi_i \gamma_{m-i} + \sigma_\varepsilon^2 \delta_m$$

where m=0,...,p, yielding p+1 equations. γm is the autocovariance function of X, σε is the standard deviation of the input noise process, and δm is the Kronecker delta function.

Because the last part of the equation is non-zero only if m=0, the equation is usually solved by representing it as a matrix for m>0, thus getting equation

$$\begin{bmatrix} \gamma_1 \\ \gamma_2 \\ \gamma_3 \\ \vdots \end{bmatrix} = \begin{bmatrix} \gamma_0 & \gamma_{-1} & \gamma_{-2} & \cdots \\ \gamma_1 & \gamma_0 & \gamma_{-1} & \cdots \\ \gamma_2 & \gamma_1 & \gamma_0 & \cdots \\ \vdots & \vdots & \vdots & \ddots \end{bmatrix} \begin{bmatrix} \varphi_1 \\ \varphi_2 \\ \varphi_3 \\ \vdots \end{bmatrix}$$

solving all φ. For m=0 have

$$\gamma_0 = \sum_{i=1}^{p} \varphi_i \gamma_{-i} + \sigma_\varepsilon^2$$

which allows us to solve σ_ε^2.

The above equations (the YuleWalker equations) provide one route to estimating the parameters of an AR(p) model, by replacing the theoretical covariances with estimated values. One way of specifying the estimated covariances is equivalent to a calculation using least squares regression of values X_t on the p previous values of the same series.

Oceanic Effects

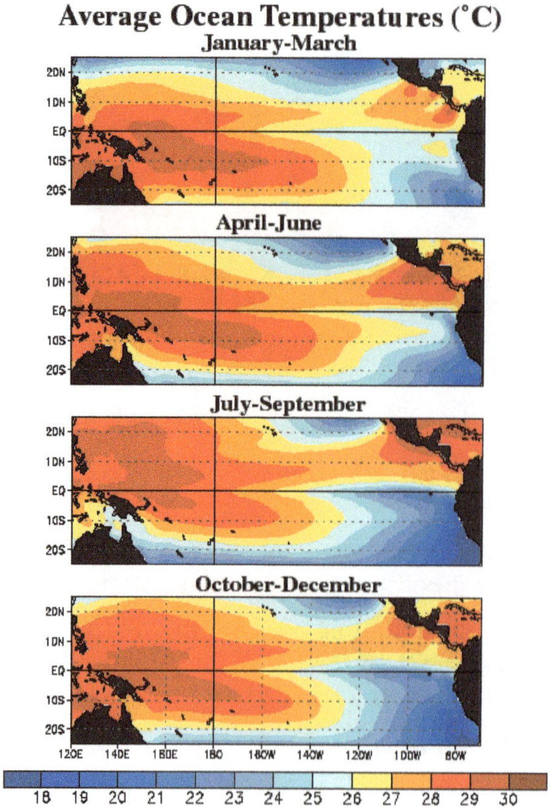

Average equatorial Pacific temperatures

The Walker Circulations of the tropical Indian, Pacific, and Atlantic basins result in westerly surface winds in Northern Summer in the first basin and easterly winds in the second and third basins. As a result, the temperature structure of the three oceans display dramatic asymmetries. The equatorial Pacific and Atlantic both have cool surface temperatures in Northern Summer in the east, while cooler surface temperatures prevail only in the western Indian Ocean. These changes in surface temperature reflect changes in the depth of the thermocline.

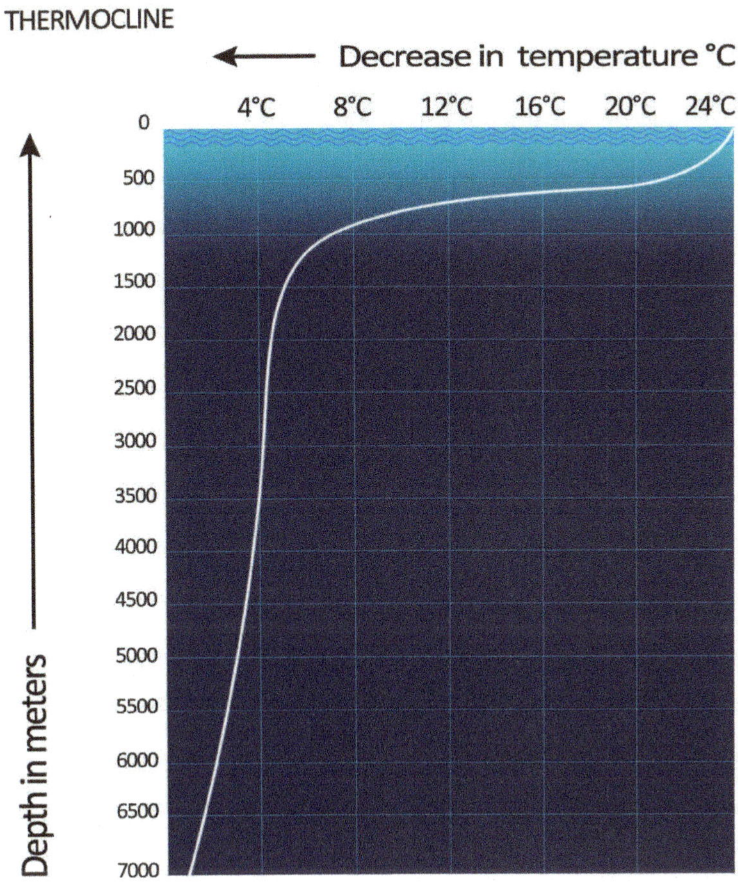

Graph showing a tropical ocean thermocline (depth vs. temperature). Note the rapid change between 100 and 1000 meters. The temperature is nearly constant after 1500 meters depth.

Changes in the Walker Circulation with time occur in conjunction with changes in surface temperature. Some of these changes are forced externally, such as the seasonal shift of the Sun into the Northern Hemisphere in summer. Other changes appear to be the result of coupled ocean-atmosphere feedback in which, for example, easterly winds cause the sea surface temperature to fall in the east, enhancing the zonal heat contrast and hence intensifying easterly winds across the basin. These anomalous easterlies induce more equatorial upwelling and raise the thermocline in the east, amplifying the initial cooling by the southerlies. This coupled ocean-atmosphere feedback was originally proposed by Bjerknes. From an oceanographic point of view, the equatorial cold tongue is caused by easterly winds. Were the earth climate symmetric about the equator, cross-equatorial wind would vanish, and the cold tongue would be much weaker and have a very different zonal structure than is observed today. The Walker cell is indirectly related to upwelling off the coasts of Peru and Ecuador. This brings nutrient-rich cold water to the surface, increasing fishing stocks.

El Niño

The Walker circulation is caused by the pressure gradient force that results from a high pressure system over the eastern Pacific Ocean, and a low pressure system over Indonesia. When the Walker circulation weakens or reverses, an El Niño results, causing the ocean surface to be warmer than average, as upwelling of cold water occurs less or not at all. An especially strong Walker circulation causes a La Niña, resulting in cooler ocean temperatures due to increased upwelling.

A scientific study published in May 2006 in the journal Nature indicates that the Walker circulation has been slowing since the mid-19th Century. The authors argue that global warming is a likely causative factor in the weakening of the wind pattern. However, a new study from The Twentieth Century Reanalysis Project shows that the Walker circulation has not been slowing (or increasing) from 1871–2008.

EL NIÑO–SOUTHERN OSCILLATION

El Niño–Southern Oscillation (ENSO) is an irregularly periodical variation in winds and sea surface temperatures over the tropical eastern Pacific Ocean, affecting much of the tropics and subtropics. The warming phase is known as El Niño and the cooling phase as La Niña. Southern Oscillation is the accompanying atmospheric component, coupled with the sea temperature change: El Niño is accompanied with high, and La Niña with low air surface pressure in the tropical western Pacific. The two periods last several months each (typically occur every few years) and their effects vary in intensity.

The two phases relate to the Walker circulation, discovered by Gilbert Walker during the early twentieth century. The Walker circulation is caused by the pressure gradient force that results from a high pressure system over the eastern Pacific Ocean, and a low pressure system over Indonesia. When the Walker circulation weakens or reverses, an El Niño results, causing the ocean surface to be warmer than average, as upwelling of cold water occurs less or not at all. An especially strong Walker circulation causes a La Niña, resulting in cooler ocean temperatures due to increased upwelling.

Mechanisms that cause the oscillation remain under study. The extremes of this climate pattern's oscillations cause extreme weather (such as floods and droughts) in many regions of the world. Developing countries dependent upon agriculture and fishing, particularly those bordering the Pacific Ocean, are the most affected.

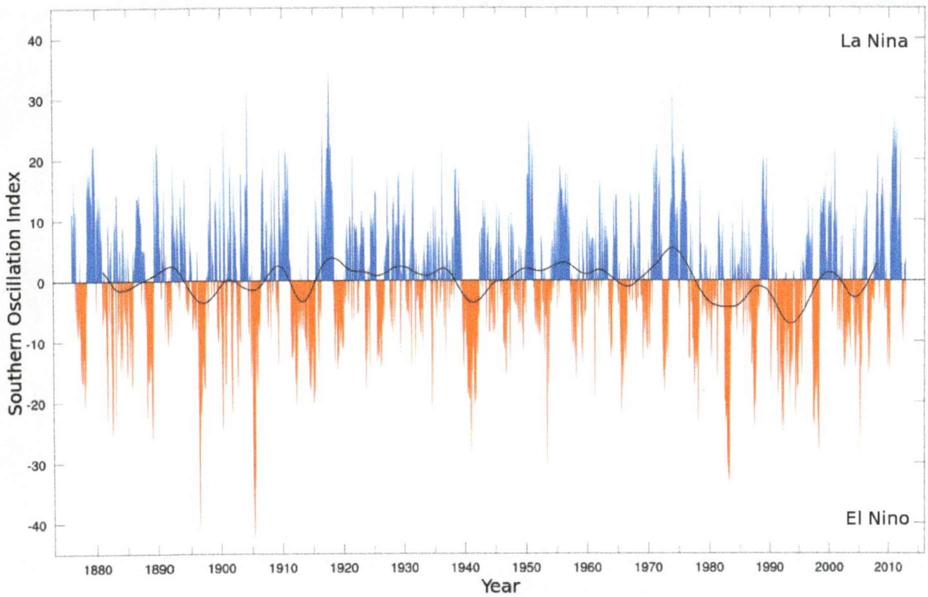

Southern Oscillation Index timeseries 1876–2012.

Concept

The El Niño–Southern Oscillation is a single climate phenomenon that periodically fluctuates between 3 phases: Neutral, La Niña or El Niño. La Niña and El Niño are opposite phases that require certain changes to take place in both the ocean and the atmosphere, before an event is declared.

Normally the northward flowing Humboldt Current brings relatively cold water from the Southern Ocean northwards along South America's west coast to the tropics, where it is enhanced by up-welling taking place along the coast of Peru. Along the equator trade winds cause the ocean currents in the eastern Pacific to draw water from the deeper ocean towards the surface, helping to keep the surface cool. Under the influence of the equatorial trade winds, this cold water flows westwards along the equator where it is slowly heated by the sun. As a direct result sea surface temperatures in the western Pacific are generally warmer, by about 8–10 °C (14–18 °F) than those in the Eastern Pacific. This warmer area of ocean is a source for convection and is associated with cloudiness and rainfall. During El Niño years the cold water weakens or disappears completely as the water in the Central and Eastern Pacific becomes as warm as the Western Pacific.

Sea Surface Temperature Oscillation

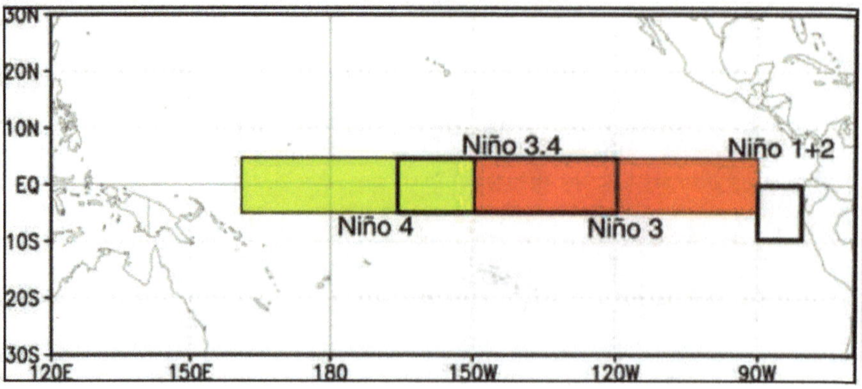

The various "Niño regions" where sea surface temperatures are monitored to determine the current ENSO phase (warm or cold)

Within the National Oceanic and Atmospheric Administration in the United States, sea surface temperatures in the Niño 3.4 region, which stretches from the 120th to 170th meridians west longitude astride the equator five degrees of latitude on either side, is monitored. This region is approximately 3,000 kilometres (1,900 mi) to the southeast of Hawaii. The most recent three-month average for the area is computed, and if the region is more than 0.5 °C (0.9 °F) above (or below) normal for that period, then an El Niño (or La Niña) is considered in progress. The United Kingdom's MetOffice also uses a several month period to determine ENSO state. When this warming or cooling occurs for only seven to nine months, it is classified as El Niño/La Niña "conditions"; when it occurs for more than that period, it is classified as El Niño/La Niña "episodes".

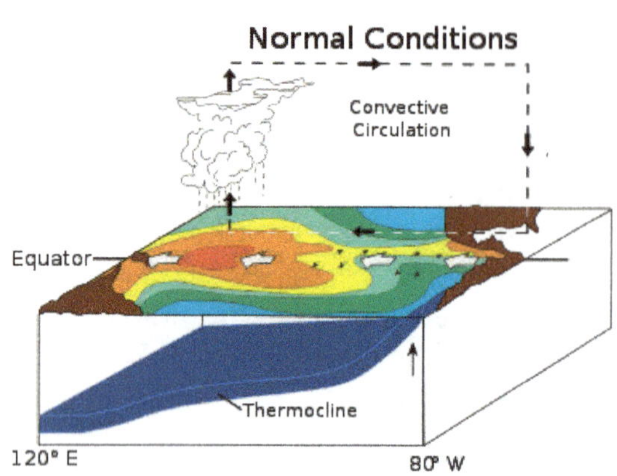

Normal Pacific pattern: Equatorial winds gather warm water pool toward the west. Cold water upwells along South American coast. (NOAA / PMEL / TAO)

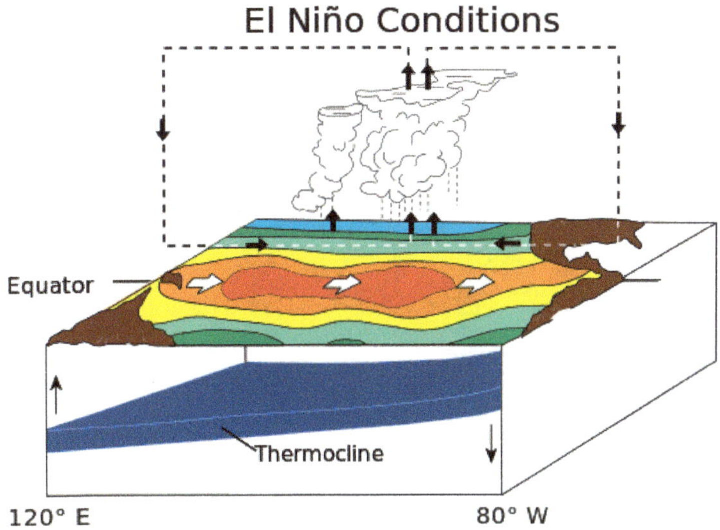

El Niño conditions: Warm water pool approaches the South American coast. The absence of cold upwelling increases warming.

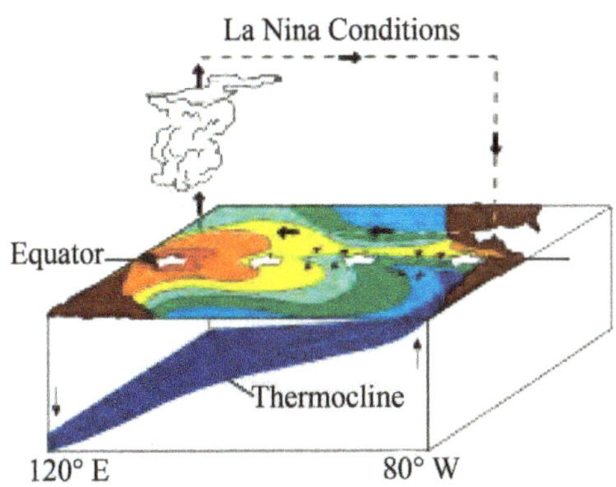

La Niña conditions: Warm water is farther west than usual.

Neutral Phase

If the temperature variation from climatology is within 0.5 °C (0.9 °F), ENSO conditions are described as neutral. Neutral conditions are the transition between warm and cold phases of ENSO. Ocean temperatures (by definition), tropical precipitation, and wind patterns are near average conditions during this phase. Close to half of all years are within neutral periods. During the neutral ENSO phase, other climate anomalies/patterns such as the sign of the North Atlantic Oscillation or the Pacific–North American teleconnection pattern exert more influence.

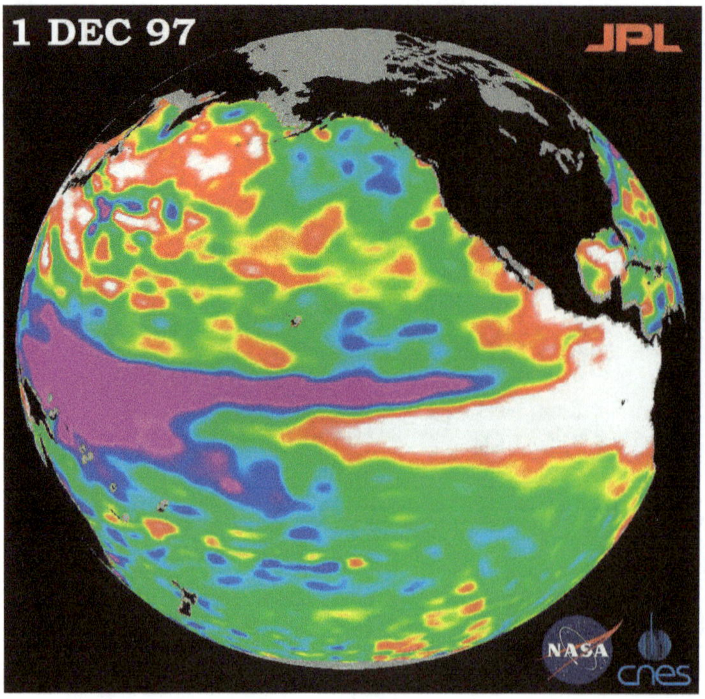

The 1997 El Niño observed by TOPEX/Poseidon

Warm Phase

When the Walker circulation weakens or reverses and the Hadley circulation strengthens an El Niño results, causing the ocean surface to be warmer than average, as upwelling of cold water occurs less or not at all offshore northwestern South America. El Niño is associated with a band of warmer than average ocean water temperatures that periodically develops off the Pacific coast of South America. *El niño* is Spanish for "the boy", and the capitalized term El Niño refers to the Christ child, Jesus, because periodic warming in the Pacific near South America is usually noticed around Christmas. It is a phase of 'El Niño–Southern Oscillation' (ENSO), which refers to variations in the temperature of the surface of the tropical eastern Pacific Ocean and in air surface pressure in the tropical western Pacific. The warm oceanic phase, El Niño, accompanies high air surface pressure in the western Pacific. Mechanisms that cause the oscillation remain under study.

Cold Phase

An especially strong Walker circulation causes a La Niña, resulting in cooler ocean temperatures in the central and eastern tropical Pacific Ocean due to increased upwelling. La Niña is a coupled ocean-atmosphere phenomenon that is the counterpart of El Niño as part of the broader El Niño Southern Oscillation climate pattern. The name La Niña originates from Spanish, meaning "the girl", analogous to El Niño meaning "the boy".

During a period of La Niña, the sea surface temperature across the equatorial eastern central Pacific will be lower than normal by 3–5 °C. In the United States, an *appearance* of La Niña happens for at least five months of La Niña conditions. However, each country and island nation has a different threshold for what constitutes a La Niña event, which is tailored to their specific interests. The Japan Meteorological Agency for example, declares that a La Niña event has started when the average five month sea surface temperature deviation for the NINO.3 region, is over 0.5 °C (0.90 °F) cooler for 6 consecutive months or longer.

Transitional Phases

Transitional phases at the onset or departure of El Niño or La Niña can also be important factors on global weather by affecting teleconnections. Significant episodes, known as Trans-Niño, are measured by the Trans-Niño index (TNI). Examples of affected short-time climate in North America include precipitation in the Northwest US and intense tornado activity in the contiguous US.

Southern Oscillation

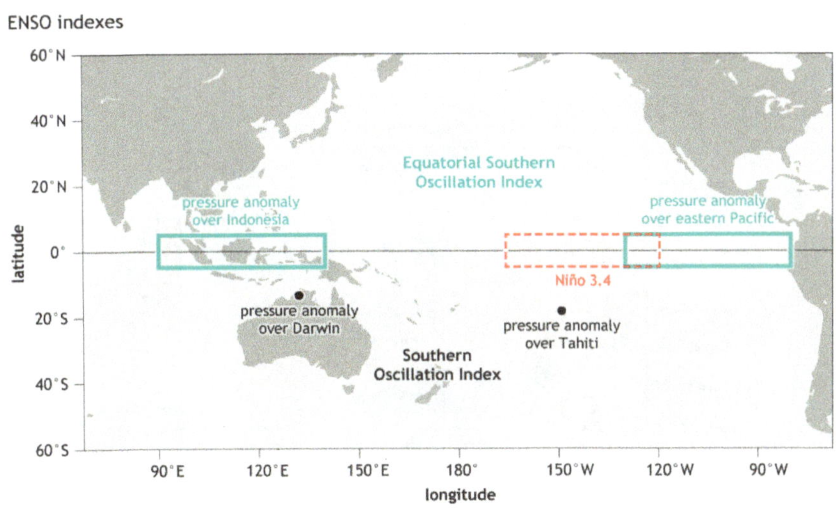

The regions where the air pressure are measured and compared to generate the Southern Oscillation Index.

The Southern Oscillation is the atmospheric component of El Niño. This component is an oscillation in surface air pressure between the tropical eastern and the western Pacific Ocean waters. The strength of the Southern Oscillation is measured by the **Southern Oscillation Index** (**SOI**). The SOI is computed from fluctuations in the surface air pressure difference between Tahiti (in the Pacific) and Darwin, Australia (on the Indian Ocean).

- **El Niño** episodes have negative SOI, meaning there is lower pressure over Tahiti and higher pressure in Darwin.

- **La Niña** episodes have positive SOI, meaning there is higher pressure in Tahiti and lower in Darwin.

Low atmospheric pressure tends to occur over warm water and high pressure occurs over cold water, in part because of deep convection over the warm water. El Niño episodes are defined as sustained warming of the central and eastern tropical Pacific Ocean, thus resulting in a decrease in the strength of the Pacific trade winds, and a reduction in rainfall over eastern and northern Australia. La Niña episodes are defined as sustained cooling of the central and eastern tropical Pacific Ocean, thus resulting in an increase in the strength of the Pacific trade winds, and the opposite effects in Australia when compared to El Niño.

Although the Southern Oscillation Index has a long station record going back to the 1800s, its reliability is limited due to the presence of both Darwin and Tahiti well south of the Equator, resulting in the surface air pressure at both locations being less directly related to ENSO. To overcome this question, a new index was created, being named Equatorial Southern Oscillation Index (EQSOI). To generate this index data, two new regions, centered on the Equator, were delimited to create a new index: The western one is located over Indonesia and the eastern one is located over equatorial Pacific, close to the South American coast. However, data on EQSOI goes back only to 1949.

Madden–Julian Oscillation

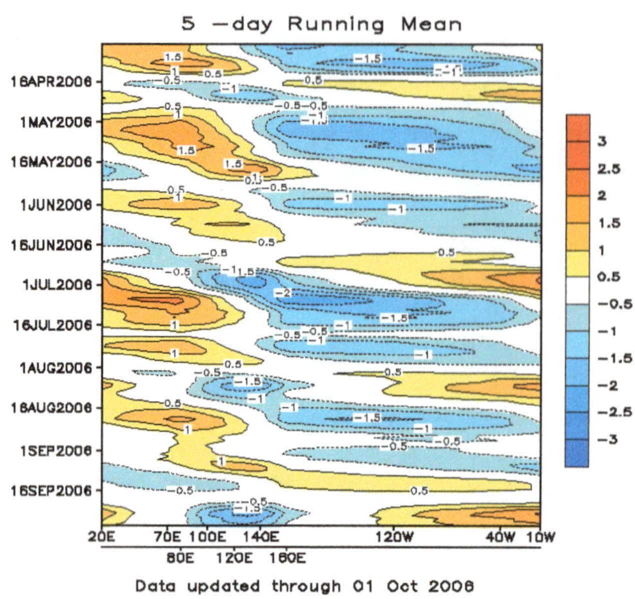

A Hovmöller diagram of the 5-day running mean of outgoing longwave radiation showing the MJO. Time increases from top to bottom in the figure, so contours that are oriented from upper-left to lower-right represent movement from west to east.

The Madden–Julian oscillation, or (MJO), is the largest element of the intraseasonal (30- to 90-day) variability in the tropical atmosphere, and was discovered by Roland Madden and Paul Julian of the National Center for Atmospheric Research (NCAR) in 1971. It is a large-scale coupling between atmospheric circulation and tropical deep convection. Rather than being a standing pattern like the El Niño Southern Oscillation (ENSO), the MJO is a traveling pattern that propagates eastward at approximately 4 to 8 m/s (9 to 18 mph), through the atmosphere above the warm parts of the Indian and Pacific oceans. This overall circulation pattern manifests itself in various ways, most clearly as anomalous rainfall. The wet phase of enhanced convection and precipitation is followed by a dry phase where thunderstorm activity is suppressed. Each cycle lasts approximately 30–60 days. Because of this pattern, The MJO is also known as the *30- to 60-day oscillation*, *30- to 60-day wave*, or *intraseasonal oscillation*.

There is strong year-to-year (interannual) variability in MJO activity, with long periods of strong activity followed by periods in which the oscillation is weak or absent. This interannual variability of the MJO is partly linked to the El Niño–Southern Oscillation (ENSO) cycle. In the Pacific, strong MJO activity is often observed 6 – 12 months prior to the onset of an El Niño episode, but is virtually absent during the maxima of some El Niño episodes, while MJO activity is typically greater during a La Niña episode. Strong events in the Madden–Julian oscillation over a series of months in the western Pacific can speed the development of an El Niño or La Niña but usually do not in themselves lead to the onset of a warm or cold ENSO event. However, observations suggest that the 1982–1983 El Niño developed rapidly during July 1982 in direct response to a Kelvin wave triggered by an MJO event during late May. Further, changes in the structure of the MJO with the seasonal cycle and ENSO might facilitate more substantial impacts of the MJO on ENSO. For example, the surface westerly winds associated with active MJO convection are stronger during advancement toward El Niño and the surface easterly winds associated with the suppressed convective phase are stronger during advancement toward La Nina.

Impacts

On Precipitation

Developing countries dependent upon agriculture and fishing, particularly those bordering the Pacific Ocean, are the most affected by ENSO. The effects of El Niño in South America are direct and strong. An El Niño is associated with warm and very wet weather months in April–October along the coasts of northern Peru and Ecuador, causing major flooding whenever the event is strong or extreme. La Niña causes a drop in sea surface temperatures over Southeast Asia and heavy rains over Malaysia, the Philippines, and Indonesia.

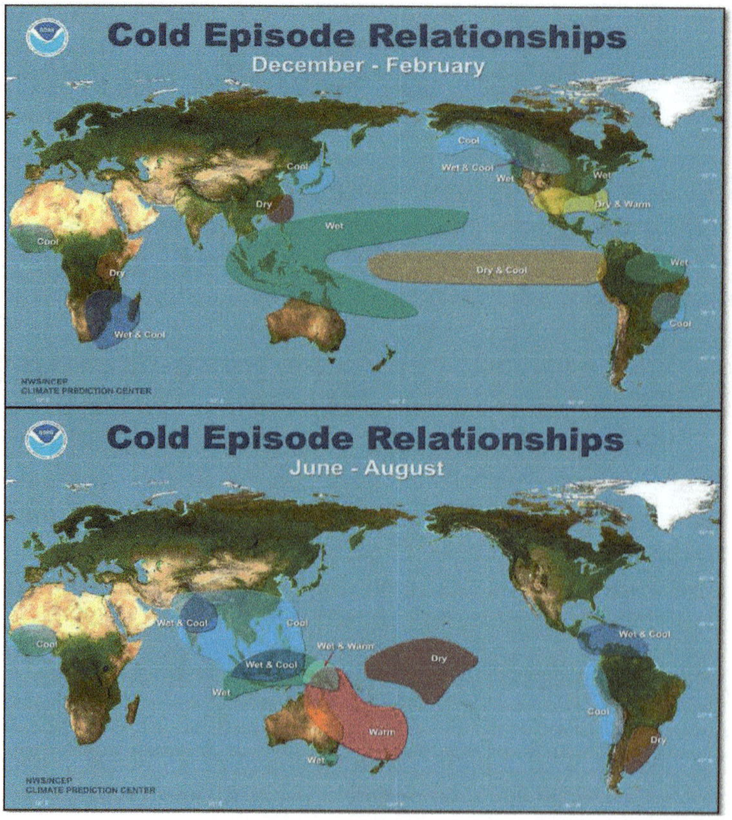

Regional impacts of La Niña.

To the north across Alaska, La Niña events lead to drier than normal conditions, while El Niño events do not have a correlation towards dry or wet conditions. During El Niño events, increased precipitation is expected in California due to a more southerly, zonal, storm track. During La Niña, increased precipitation is diverted into the Pacific Northwest due to a more northerly storm track. During La Niña events, the storm track shifts far enough northward to bring wetter than normal winter conditions (in the form of increased snowfall) to the Midwestern states, as well as hot and dry summers. During the El Niño portion of ENSO, increased precipitation falls along the Gulf coast and Southeast due to a stronger than normal, and more southerly, polar jet stream. In the late winter and spring during El Niño events, drier than average conditions can be expected in Hawaii. On Guam during El Niño years, dry season precipitation averages below normal. However, the threat of a tropical cyclone is over triple what is normal during El Niño years, so extreme shorter duration rainfall events are possible. On American Samoa during El Niño events, precipitation averages about 10 percent above normal, while La Niña events lead to precipitation amounts which average close to 10 percent below normal. ENSO is linked to rainfall over Puerto Rico. During an El Niño, snowfall is greater than average across the southern Rockies and Sierra Nevada mountain range, and is well-below normal across the Upper Midwest and Great Lakes states. During a La Niña, snowfall is above normal across the Pacific Northwest and western Great Lakes.

On Tehuantepecers

The synoptic condition for the Tehuantepecer, a violent mountain-gap wind in between the mountains of Mexico and Guatemala, is associated with high-pressure system forming in Sierra Madre of Mexico in the wake of an advancing cold front, which causes winds to accelerate through the Isthmus of Tehuantepec. Tehuantepecers primarily occur during the cold season months for the region in the wake of cold fronts, between October and February, with a summer maximum in July caused by the westward extension of the Azores-Bermuda high pressure system. Wind magnitude is greater during El Niño years than during La Niña years, due to the more frequent cold frontal incursions during El Niño winters. Tehuantepec winds reach 20 knots (40 km/h) to 45 knots (80 km/h), and on rare occasions 100 knots (200 km/h). The wind's direction is from the north to north-northeast. It leads to a localized acceleration of the trade winds in the region, and can enhance thunderstorm activity when it interacts with the Intertropical Convergence Zone. The effects can last from a few hours to six days.

On Global Warming

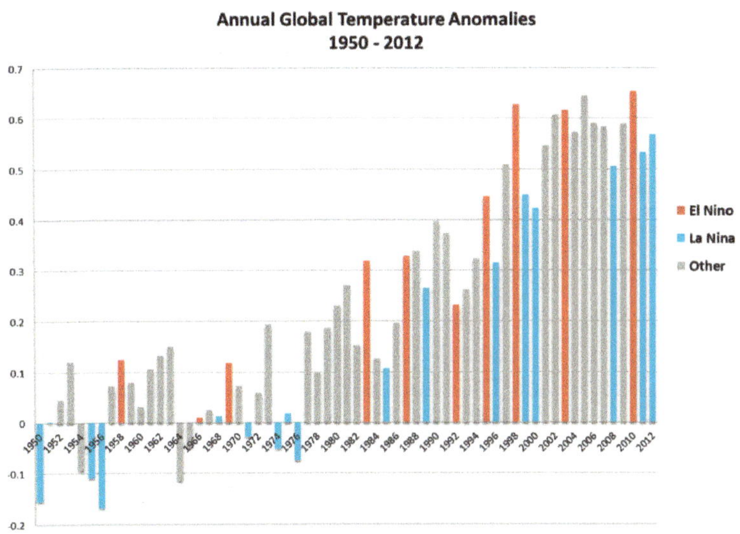

NOAA graph of Global Annual Temperature Anomalies 1950–2012, showing ENSO

El Niño events cause short-term (approximately 1 year in length) spikes in global average surface temperature while La Niña events cause short term cooling. Therefore, the relative frequency of El Niño compared to La Niña events can affect global temperature trends on decadal timescales. Over the last several decades, the number of El Niño events increased, and the number of La Niña events decreased, although observation of ENSO for much longer is needed to detect robust changes.

The studies of historical data show the recent El Niño variation is most likely linked to global warming. For example, one of the most recent results, even after subtracting the positive influence of decadal variation, is shown to be possibly present in the ENSO trend, the amplitude of the ENSO variability in the observed data still increases, by as much as 60% in the last 50 years.

The exact changes happening to ENSO in the future is uncertain: Different models make different predictions. It may be that the observed phenomenon of more frequent and stronger El Niño events occurs only in the initial phase of the global warming, and then (e.g., after the lower layers of the ocean get warmer, as well), El Niño will become weaker than it was. It may also be that the stabilizing and destabilizing forces influencing the phenomenon will eventually compensate for each other. More research is needed to provide a better answer to that question. The ENSO is considered to be a potential tipping element in Earth's climate and, under the global warming, can enhance or alternate regional climate extreme events through a strengthened teleconnection. For example, an increase in the frequency and magnitude of El Niño events have triggered warmer than usual temperatures over the Indian Ocean, by modulating the Walker circulation. This has resulted in a rapid warming of the Indian Ocean, and consequently a weakening of the Asian Monsoon.

On Coral Bleaching

Following the El Nino event in 1997 – 1998, the largest recorded to date, the Pacific Marine Environmental Laboratory attributes the first large-scale water bleaching event to the warming waters. This process, called reef symbiosis, results to warming ocean water temperatures.

ENSO Diversity

The traditional ENSO (El Niño Southern Oscillation), also called Eastern Pacific (EP) ENSO, involves temperature anomalies in the eastern pacific. However, in the 1990s and 2000s, nontraditional ENSO conditions were observed, in which the usual place of the temperature anomaly (Niño 1 and 2) is not affected, but an anomaly arises in the central Pacific (Niño 3.4). The phenomenon is called Central Pacific (CP) ENSO, "dateline" ENSO (because the anomaly arises near the dateline), or ENSO "Modoki" (Modoki is Japanese for "similar, but different"). There are flavors of ENSO additional to EP and CP types and some scientists argue that ENSO exists as a continuum often with hybrid types.

The effects of the CP ENSO are different from those of the traditional EP ENSO. The El Niño Modoki leads to more hurricanes more frequently making landfall in the Atlantic. La Niña Modoki leads to a rainfall increase over northwestern Australia and northern Murray-Darling basin, rather than over the east as in a conventional La Niña. Also, La Niña Modoki increases the frequency of cyclonic storms over Bay of Bengal, but decreases the occurrence of severe storms in the Indian Ocean.

The recent discovery of ENSO Modoki has some scientists believing it to be linked to global warming. However, comprehensive satellite data go back only to 1979. More research must be done to find the correlation and study past El Niño episodes. More generally, there is no scientific consensus on how/if climate change may affect ENSO.

There is also a scientific debate on the very existence of this «new» ENSO. Indeed, a number of studies dispute the reality of this statistical distinction or its increasing occurrence, or both, either arguing the reliable record is too short to detect such a distinction, finding no distinction or trend using other statistical approaches, or that other types should be distinguished, such as standard and extreme ENSO. Following the asymmetric nature of the warm and cold phases of ENSO, some studies could not identify such distinctions for La Niña, both in observations and in the climate models, but some sources indicate that there is a variation on La Niña with cooler waters on central Pacific and average or warmer water temperatures on both eastern and western Pacific, also showing eastern Pacific Ocean currents going to the opposite direction compared to the currents in traditional La Niñas.

REFERENCES

1. Junling Huang and Michael B. McElroy (2014). "Contributions of the Hadley and Ferrel Circulations to the Energetics of the Atmosphere over the Past 32 Years". Journal of Climate. 27 (7): 2656–2666. Bibcode:2014JCli...27.2656H. doi:10.1175/jcli-d-13-00538.1.
2. "The physical environment of the Antarctic". British Antarctic Survey (BAS).
3. "Regional climate variation and weather". RGS-IBG in partnership with BAS. Archived from the original on 2015-03-06.
4. "Welcome to the Coldest Town on Earth". Scientific American. 2008.
5. Yochanan Kushnir (2000). "The Climate System: General Circulation and Climate Zones". Retrieved 13 March 2012.
6. "Envisat watches for La Nina". BNSC. 2006-03-03. Archived from the original on 2008-04-24. Retrieved 2007-07-26.
7. "The Tropical Atmosphere Ocean Array: Gathering Data to Predict El Niño". Celebrating 200 Years. NOAA. 2007-01-08. Retrieved 2007-07-26.
8. "Ocean Surface Topography". Oceanography 101. JPL, NASA. 2006-07-05. Archived from the original on April 14, 2009. Retrieved 2007-07-26.
9. "ANNUAL SEA LEVEL DATA SUMMARY REPORT JULY 2005 – JUNE 2006" (pdf). THE AUSTRALIAN BASELINE SEA LEVEL MONITORING PROJECT. Bureau of Meteorology. Retrieved 2007-07-26.
10. Climate Prediction Center (2005-12-19). "Frequently Asked Questions about El Niño and La Niña". National Centers for Environmental Prediction. Retrieved 2009-07-17.
11. Trenberth, K.E., P.D. Jones, P. Ambenje, R. Bojariu , D. Easterling, A. Klein Tank, D. Parker, F. Rahimzadeh, J.A. Renwick, M. Rusticucci, B. Soden and P. Zhai. "Observations: Surface and Atmospheric Climate Change". In Solomon, S., D. Qin, M. Manning, Z. Chen, M. Marquis, K.B. Averyt, M. Tignor and H.L. Miller. Climate Change 2007: The Physical Science Basis. Contribution of Working Group I to the Fourth Assessment Report of the Intergovernmental Panel on Climate Change. Cambridge, UK: Cambridge University Press. pp.

235–336.

12. "El Niño, La Niña and the Southern Oscillation". MetOffice. Retrieved 2015-08-18.
13. L'Heureux, Michelle (5 May 2014). "What is the El Niño–Southern Oscillation (ENSO) in a nutshell?". ENSO Blog. Archived from the original on 10 April 2016.
14. "El Niño, La Niña and Australia's Climate" (PDF). Australian Bureau of Meteorology. 6 May 2005. Archived(PDF) from the original on 10 April 2016.
15. "El Niño Southern Oscillation (ENSO)". Australian Bureau of Meteorology. 2 April 2008. Archived (PDF) from the original on 10 April 2016.
16. Bureau of Meteorology. "The Walker Circulation". Commonwealth of Australia. Retrieved 2014-07-01.
17. Zelle, Hein, Gerrian Appledoorn, Gerritt Burgers, and Gert Jan Van Oldenborgh (March 2004). "Relationship Between Sea Surface Temperature and Thermocline Depth in the Eastern Equatorial Pacific". Journal of Physical Oceanography. American Meteorological Society. 34 (3): 643–655. Bibcode:2004JPO....34..643Z. doi:10.1175/2523.1.
18. "Ocean-atmosphere interaction in the making of the Walker circulation and equatorial cold tongue". inist.fr. Retrieved 20 October 2015.
19. Jennings, S., Kaiser, M.J., Reynolds, J.D. (2001) "Marine Fisheries Ecology." Oxford: Blackwell Science Ltd. ISBN 0-632-05098-5
20. Pidwirny, Michael (2006-02-02). "Chapter 7: Introduction to the Atmosphere". Fundamentals of Physical Geography. physicalgeography.net. Retrieved 2006-12-30.
21. "Envisat watches for La Niña". BNSC via the Internet Wayback Machine. 2011-01-09. Archived from the originalon 2008-04-24. Retrieved 2007-07-26.
22. "The Tropical Atmosphere Ocean Array: Gathering Data to Predict El Niño". Celebrating 200 Years. NOAA. 2007-01-08. Retrieved 2007-07-26.
23. "Ocean Surface Topography". Oceanography 101. JPL. 2006-07-05. Archived from the original on 2009-04-14. Retrieved 2007-07-26."Annual Sea Level Data Summary Report July 2005 – June 2006" (PDF). The Australian Baseline Sea Level Monitoring Project. Bureau of Meteorology. Archived from the original (PDF) on 2007-08-07. Retrieved 2007-07-26.
24. Climate Prediction Center (2014-06-30). "ENSO: Recent Evolution, Current Status and Predictions" (PDF). National Oceanic and Atmospheric Administration. pp. 5, 19–20. Retrieved 2014-06-30.
25. MetOffice (2012-10-11). "El Niño, La Niña and the Southern Oscillation". United Kingdom. Retrieved 2014-06-30.
26. National Climatic Data Center (June 2009). "El Niño / Southern Oscillation (ENSO) June 2009". National Oceanic and Atmospheric Administration. Retrieved 2009-07-26.

27. Climate Prediction Center Internet Team (2012-04-26). "Frequently Asked Questions about El Niño and La Niña". National Oceanic and Atmospheric Administration. Retrieved 2014-06-30.

28. International Research Institute for Climate and Society (February 2002). "More Technical ENSO Comment". Columbia University. Retrieved 2014-06-30.

29. State Climate Office of North Carolina. "Global Patterns – El Niño-Southern Oscillation (ENSO)". North Carolina State University. Retrieved 2014-06-30.

30. Intergovernmental Panel on Climate Change (2007). "Climate Change 2007: Working Group I: The Physical Science Basis: 3.7 Changes in the Tropics and Subtropics, and the Monsoons". World Meteorological Organization. Retrieved 2014-07-01.

31. "El Niño Information". California Department of Fish and Game, Marine Region.

32. Trenberth, K.E., P.D. Jones, P. Ambenje, R. Bojariu , D. Easterling, A. Klein Tank, D. Parker, F. Rahimzadeh, J.A. Renwick, M. Rusticucci, B. Soden and P. Zhai. "Observations: Surface and Atmospheric Climate Change". In Solomon, S., D. Qin, M. Manning, Z. Chen, M. Marquis, K.B. Averyt, M. Tignor and H.L. Miller. Climate Change 2007: The Physical Science Basis. Contribution of Working Group I to the Fourth Assessment Report of the Intergovernmental Panel on Climate Change. Cambridge, UK: Cambridge University Press. pp. 235–336.

33. Tropical Atmosphere Ocean project (2008-03-24). "What is La Niña?". Pacific Marine Environmental Laboratory. Retrieved 2009-07-17.

34. Becker, Emily (4 December 2014). "December's ENSO Update: Close, but no cigar". ENSO Blog. Archived from the original on 4 April 2016.

35. "Historical El Niño and La Niña Events". Japan Meteorological Agency. Retrieved April 4, 2016.

36. Trenberth, Kevin E.; D. P. Stepaniak (2001). "Indices of El Niño Evolution". J. Climate. 14 (8): 1697–701. Bibcode:2001JCli...14.1697T. doi:10.1175/1520-0442(2001)014<1697:LIOENO>2.0.CO;2. ISSN 1520-0442.

37. Kennedy, Adam M.; D. C. Garen; R. W. Koch (2009). "The association between climate teleconnection indices and Upper Klamath seasonal streamflow: Trans-Niño Index". Hydrol. Process. 23 (7): 973–84. Bibcode:2009HyPr...23..973K. doi:10.1002/hyp.7200.

38. Lee, Sang-Ki; R. Atlas; D. Enfield; C. Wang; H. Liu (2013). "Is there an optimal ENSO pattern that enhances large-scale atmospheric processes conducive to tornado outbreaks in the U.S?". J. Climate. 26 (5): 1626–1642. Bibcode:2013JCli...26.1626L. doi:10.1175/JCLI-D-12-00128.1.

39. "Climate glossary — Southern Oscilliation Index (SOI)". Bureau of Meteorology (Australia). 2002-04-03. Retrieved 2009-12-31.

40. Barnston, Anthony (2015-01-29). "Why are there so many ENSO indexes, instead of just one?". NOAA. Retrieved 2015-08-14.

41. International Research Institute for Climate and Society. "Southern Oscillation Index (SOI) and Equatorial SOI". Columbia University. Retrieved 2015-08-14.
42. Zhang, Chidong (2005). "Madden–Julian Oscillation". Retrieved 22 February 2012.
43. "Madden–Julian oscillation forecast research". University of East Anglia. Retrieved 22 February 2012.
44. Takmeng Wong; G. Louis Smith; T. Dale Bess. "P1.38 Radiative Energy Budget of African Monsoons: NASA Ceres Observations Versus NOAA NCEP Reanalysis 2 Data"(PDF). Retrieved 2009-11-06.
45. Gottschalck, Jon; Wayne Higgins (2008-02-16). "Madden Julian Oscillation Impacts" (PDF). Climate Prediction Center. Retrieved 2009-07-17.
46. Roundy, Paul E.; Kiladis, George N. (2007). "An Error Occurred Setting Your User Cookie". ametsoc.org. p. 4341. doi:10.1175/JCLI4249.1. Retrieved 20 October 2015.
47. Roundy, Paul E.; Kravitz, Joseph R. (2009). "An Error Occurred Setting Your User Cookie". ametsoc.org. p. 381. doi:10.1175/2008JCLI2389.1. Retrieved 20 October 2015.
48. "Atmospheric Consequences of El Niño". University of Illinois. Retrieved 2010-05-31.
49. Hong, Lynda (2008-03-13). "Recent heavy rain not caused by global warming". Channel NewsAsia. Retrieved 2008-06-22.
50. Monteverdi, John and Jan Null. WESTERN REGION TECHNICAL ATTACHMENT NO. 97-37 NOVEMBER 21, 1997: El Niño and California Precipitation. Retrieved on 2008-02-28.
51. Mantua, Nathan. La Niña Impacts in the Pacific Northwest.Retrieved on 2008-02-29.
52. Reuters. La Nina could mean dry summer in Midwest and Plains. Retrieved on 2008-02-29.
53. Climate Prediction Center. El Niño (ENSO) Related Rainfall Patterns Over the Tropical Pacific. Retrieved on 2008-02-28.
54. Chu, Pao-Shin. Hawaii Rainfall Anomalies and El Niño.Retrieved on 2008-03-19.
55. Pacific ENSO Applications Climate Center. Pacific ENSO Update: 4th Quarter, 2006. Vol. 12 No. 4. Retrieved on 2008-03-19.
56. Pacific ENSO Applications Climate Center. RAINFALL VARIATIONS DURING ENSO. Retrieved on 2008-03-19.
57. San Juan, Puerto Rico Weather Forecast Office (2010-09-02). "The Local Impacts of ENSO across the Northeastern Caribbean". National Weather Service Southern Region Headquarters. Retrieved 2014-07-01.
58. Climate Prediction Center. ENSO Impacts on United States Winter Precipitation and Temperature. Retrieved on 2008-04-16.
59. Romero-Centeno, Rosario, Jorge Zavala-Hidalgo, Artemio Gallegos, and James

J. O'Brien (August 2003). "Isthmus of Tehuantepec wind climatology and ENSO signal". Journal of Climate. 16 (15): 2628–2639. Bibcode:2003JCli...16.2628R. doi:10.1175/1520-0442(2003)016<2628:iotwca>2.0.co;2.

60. American Meteorological Society (2012-01-26). "Tehuantepecer". Glossary of Meteorology. Retrieved 2013-05-16.
61. Fett, Bob (2002-12-09). "World Wind Regimes – Central America Gap Wind Tutorial". United States Naval Research Laboratory Monterey, Marine Meteorology Division. Retrieved 2013-05-16.
62. Arnerich, Paul A. "Tehuantepecer Winds of the West Coast of Mexico". Mariners Weather Log. National Oceanic and Atmospheric Administration. 15 (2): 63–67.
63. Brown, Patrick T.; Li, Wenhong; Xie, Shang-Ping (2015-01-27). "Regions of significant influence on unforced global mean surface air temperature variability in climate models". Journal of Geophysical Research: Atmospheres. 120 (2): 2014JD022576. Bibcode:2015JGRD..120..480B. doi:10.1002/2014JD022576. ISSN 2169-8996.
64. Trenberth, Kevin E.; Fasullo, John T. (2013-12-01). "An apparent hiatus in global warming?". Earth's Future. 1 (1): 19–32. Bibcode:2013EaFut...1...19T. doi:10.1002/2013EF000165. ISSN 2328-4277.
65. Trenberth, Kevin E.; Hoar, Timothy J. (January 1996). "The 1990–1995 El Niño–Southern Oscillation event: Longest on record". Geophysical Research Letters. 23 (1): 57–60. Bibcode:1996GeoRL..23...57T. doi:10.1029/95GL03602.
66. Wittenberg, A.T. (2009). "Are historical records sufficient to constrain ENSO simulations?". Geophys. Res. Lett. 36 (12): L12702. Bibcode:2009GeoRL..3612702W. doi:10.1029/2009GL038710.
67. Fedorov, Alexey V.; Philander, S. George (2000). "Is El Niño Changing?". Science. 288 (5473): 1997–2002. Bibcode:2000Sci...288.1997F. doi:10.1126/science.288.5473.1997. PMID 10856205.
68. Zhang, Qiong; Guan, Yue; Yang, Haijun (2008). "ENSO Amplitude Change in Observation and Coupled Models". Advances in Atmospheric Sciences. 25 (3): 331–6. Bibcode:2008AdAtS..25..361Z. doi:10.1007/s00376-008-0361-5.
69. Collins, M.; An, S-I; Cai, W.; Ganachaud, A.; Guilyardi, E.; Jin, F-F; Jochum, M.; Lengaigne, M.; Power, S.; Timmermann, A.; Vecchi, G.; Wittenberg, A. (2010). "The impact of global warming on the tropical Pacific Ocean and El Niño". Nature Geosci. 3 (6): 391–7. Bibcode:2010NatGe...3..391C. doi:10.1038/ngeo868.
70. Merryfield, William J. (2006). "Changes to ENSO under CO2Doubling in a Multimodel Ensemble" (PDF). Journal of Climate. 19 (16): 4009–27. Bibcode:2006JCli...19.4009M. doi:10.1175/JCLI3834.1.
71. Guilyardi, E.; Wittenberg, Andrew; Fedorov, Alexey; Collins, Mat; Wang, Chunzai; Capotondi, Antonietta; Van Oldenborgh, Geert Jan; Stockdale, Tim (2009). "Understanding El Nino in Ocean-Atmosphere General Circulation Models: Progress and Challenges". American Meteorological Society. 90 (3): 325–340. Bibcode:2009BAMS...90..325G. doi:10.1175/2008BAMS2387.1.

72. Meehl, G. A.; Teng, H.; Branstator, G. (2006). "Future changes of El Niño in two global coupled climate models". Climate Dynamics. 26 (6): 549–566. Bibcode:2006ClDy...26..549M. doi:10.1007/s00382-005-0098-0.
73. Philip, S.; Van Oldenborgh, G. J. (2006). "Shifts in ENSO coupling processes under global warming". Geophysical Research Letters. 33 (11): L11704. Bibcode:2006GeoRL..3311704P. doi:10.1029/2006GL026196.
74. Lenton, T. M.; Held, H.; Kriegler, E.; Hall, J. W.; Lucht, W.; Rahmstorf, S.; Schellnhuber, H. J. (2008). "Inaugural Article: Tipping elements in the Earth's climate system". Proceedings of the National Academy of Sciences. 105 (6): 1786–1793. Bibcode:2008PNAS..105.1786L. doi:10.1073/pnas.0705414105.
75. Wang et al. (2015) "Role of the strengthened El Niño teleconnection in the May 2015 floods over the southern Great Plains" http://onlinelibrary.wiley.com/doi/10.1002/2015GL065211/full
76. Roxy, Mathew Koll; Ritika, Kapoor; Terray, Pascal; Masson, Sébastien (2014-09-11). "The Curious Case of Indian Ocean Warming". Journal of Climate. 27 (22): 8501–8509. Bibcode:2014JCli...27.8501R. doi:10.1175/JCLI-D-14-00471.1. ISSN 0894-8755.
77. Roxy, Mathew Koll; Ritika, Kapoor; Terray, Pascal; Murtugudde, Raghu; Ashok, Karumuri; Goswami, B. N. (2015-06-16). "Drying of Indian subcontinent by rapid Indian Ocean warming and a weakening land-sea thermal gradient". Nature Communications. 6: 7423. Bibcode:2015NatCo...6E7423R. doi:10.1038/ncomms8423. PMID 26077934.
78. "The 1982–83 El Nino". www.fcst-office.com. Retrieved 2016-11-12.
79. "FAQs | El Nino Theme Page – A comprehensive Resource". www.pmel.noaa.gov. Retrieved 2016-11-12.
80. "El Nino coral damage 'could be worst ever'". Mail Online. Retrieved 2016-11-12.
81. Kao, Hsun-Ying; Jin-Yi Yu (2009). "Contrasting Eastern-Pacific and Central-Pacific Types of ENSO". J. Climate. 22 (3): 615–632. Bibcode:2009JCli...22..615K. doi:10.1175/2008JCLI2309.1.
82. Larkin, N. K.; Harrison, D. E. (2005). "On the definition of El Niño and associated seasonal average U.S. Weather anomalies". Geophysical Research Letters. 32 (13): L13705. Bibcode:2005GeoRL..3213705L. doi:10.1029/2005GL022738.
83. Yuan Yuan; HongMing Yan (2012). "Different types of La Niña events and different responses of the tropical atmosphere". Chinese Science Bulletin. 58 (3): 406–415. doi:10.1007/s11434-012-5423-5.
84. Cai, W.; Cowan, T. (2009). "La Niña Modoki impacts Australia autumn rainfall variability". Geophysical Research Letters. 36 (12): L12805. Bibcode:2009GeoRL..3612805C. doi:10.1029/2009GL037885. ISSN 0094-8276.
85. Johnson, Nathaniel C. (2013). "How Many ENSO Flavors Can We Distinguish?". J. Climate. 26 (13): 4816–27. Bibcode:2013JCli...26.4816J. doi:10.1175/JCLI-D-12-00649.1.
86. Hye-Mi Kim; Peter J. Webster; Judith A. Curry (2009). "Impact of Shifting

Patterns of Pacific Ocean Warming on North Atlantic Tropical Cyclones". Science. 335 (5936): 77–80. Bibcode:2009Sci...325...77K. doi:10.1126/science.1174062. PMID 19574388.

87. Cai, W.; Cowan, T. (2009). "La Niña Modoki impacts Australia autumn rainfall variability". Geophysical Research Letters. 36 (12): L12805. Bibcode:2009GeoRL..3612805C. doi:10.1029/2009GL037885.
88. M R Ramesh Kumar (2014-04-23). "El Nino, La Nina and the Indian sub-continent". Society for Environmental Communications. Retrieved 2014-07-25.
89. Yeh, Sang-Wook; Kug, Jong-Seong; Dewitte, Boris; Kwon, Min-Ho; Kirtman, Ben P.; Jin, Fei-Fei (September 2009). "El Niño in a changing climate". Nature. 461 (7263): 511–4. Bibcode:2009Natur.461..511Y. doi:10.1038/nature08316. PMID 19779449.
90. Nicholls, N. (2008). "Recent trends in the seasonal and temporal behaviour of the El Niño Southern Oscillation". Geophys. Res. Lett. 35 (19): L19703. Bibcode:2008GeoRL..3519703N. doi:10.1029/2008GL034499.
91. McPhaden, M.J.; Lee, T.; McClurg, D. (2011). "El Niño and its relationship to changing background conditions in the tropical Pacific Ocean". Geophys. Res. Lett. 38 (15): L15709. Bibcode:2011GeoRL..3815709M. doi:10.1029/2011GL048275.
92. Giese, B.S.; Ray, S. (2011). "El Niño variability in simple ocean data assimilation (SODA), 1871–2008". J. Geophys. Res. 116: C02024. Bibcode:2011JGRC..116.2024G. doi:10.1029/2010JC006695.
93. Newman, M.; Shin, S.-I.; Alexander, M.A. (2011). "Natural variation in ENSO flavors". Geophys. Res. Lett. 38 (14): L14705. Bibcode:2011GeoRL..3814705N. doi:10.1029/2011GL047658.
94. Yeh, S.-W.; Kirtman, B.P.; Kug, J.-S.; Park, W.; Latif, M. (2011). "Natural variability of the central Pacific El Niño event on multi-centennial timescales". Geophys. Res. Lett. 38 (2): L02704. Bibcode:2011GeoRL..38.2704Y. doi:10.1029/2010GL045886.
95. Hanna Na; Bong-Geun Jang; Won-Moon Choi; Kwang-Yul Kim (2011). "Statistical simulations of the future 50-year statistics of cold-tongue El Niño and warm-pool El Niño". Asia-Pacific J. Atmos. Sci. 47 (3): 223–233. Bibcode:2011APJAS..47..223N. doi:10.1007/s13143-011-0011-1.
96. L'Heureux, M.; Collins, D.; Hu, Z.-Z. (2012). "Linear trends in sea surface temperature of the tropical Pacific Ocean and implications for the El Niño-Southern Oscillation". Climate Dynamics. 40 (5–6): 1–14. Bibcode:2013ClDy...40.1223L. doi:10.1007/s00382-012-1331-2.
97. Lengaigne, M.; Vecchi, G. (2010). "Contrasting the termination of moderate and extreme El Niño events in coupled general circulation models". Climate Dynamics. 35 (2–3): 299–313. Bibcode:2010ClDy...35..299L. doi:10.1007/s00382-009-0562-3.
98. Takahashi, K.; Montecinos, A.; Goubanova, K.; Dewitte, B. (2011). "ENSO regimes: Reinterpreting the canonical and Modoki El Niño". Geophys. Res. Lett. 38 (10): L10704. Bibcode:2011GeoRL..3810704T. doi:10.1029/2011GL047364.

99. Kug, J.-S.; Jin, F.-F.; An, S.-I. (2009). "Two types of El Niño events: Cold Tongue El Niño and Warm Pool El Niño". J. Climate. 22 (6): 1499–1515. Bibcode:2009JCli...22.1499K. doi:10.1175/2008JCLI2624.1.

100. Shinoda, Toshiaki; Hurlburt, Harley E.; Metzger, E. Joseph (2011). "Anomalous tropical ocean circulation associated with La Niña Modoki". Journal of Geophysical Research: Oceans. 115 (12): C12001. Bibcode:2011JGRC..11612001S. doi:10.1029/2011JC007304.

101. Junling Huang; Michael B. McElroy (2014). "Contributions of the Hadley and Ferrel Circulations to the Energetics of the Atmosphere over the Past 32 Years". Journal of Climate. 27 (7): 2656–2666. Bibcode:2014JCli...27.2656H. doi:10.1175/jcli-d-13-00538.1.

102. Anders Persson (2006). "Hadley's Principle: Understanding and Misunderstanding the Trade Winds" (PDF). History of Meteorology. 3: 17–42.

103. Xiao-Wei Quan; Henry F. Diaz; Martin P. Hoerling (2004). "Changes in the Tropical Hadley Cell since 1950". In Henry F. Diaz; Raymond S. Bradley. The Hadley Circulation: Present, Past, and Future. Advances in Global Change Research. 21. Springer Netherlands. pp. 85–120. doi:10.1007/978-1-4020-2944-8. ISBN 978-1-4020-2943-1. Preprint at 'Change of the Tropical Hadley Cell Since 1950', NOAA-CIRES Climate Diagnostic Center (2004) (PDF file 2.9 MB)

104. Dargan M.W. Frierson; Jian Lu; Gang Chen (2007). "Width of the Hadley cell in simple and comprehensive general circulation models" (PDF). Geophysical Research Letters. 34 (18): L18804. Bibcode:2007GeoRL..3418804F. doi:10.1029/2007GL031115.

105. Dian J. Seidel; Qian Fu; William J. Randel; Thomas J. Reichler (2007). "Widening of the tropical belt in a changing climate". Nature Geoscience. 1 (1): 21–4. Bibcode:2008NatGe...1...21S. doi:10.1038/ngeo.2007.38.

106. Celeste M. Johanson; Qiang Fu (2009). "Hadley Cell Widening: Model Simulations versus Observations" (PDF). Journal of Climate. 22 (10): 2713–25. Bibcode:2009JCli...22.2713J. doi:10.1175/2008JCLI2620.1.

107. van Geel B.,van der Plicht, J.,Kilian, M.R., (1998). "The sharp rise of 14C ca. 800 cal BP:possible causes, related climatic teleconnections and the impact on human environments.". Radiocarbon. 40 (1): 535–550.

108. van Geel B., Renssen, H., (1998). "Abrupt climate change around 2650 BP in North-West Europe:evidence for climatic teleconnections and a tentative explanation". In Issar, A.S.,Brown, N. Water, Environment and Society in Times of Climatic Change. Kluwer Academic Publishers, Dordrecht. pp. 21–41.

CHAPTER 7

Meteorological Phenomena

Chapter Contents

- **Introduction**
- **Acid Rain**
- **Air Mass**
- **Polar Vortex**
- **Ball Lightning**

INTRODUCTION

A meteorological phenomenon is a weather event that can be explained by the principles of meteorology. Such events include:

ACID RAIN

Acid rain is a rain or any other form of precipitation that is unusually acidic, meaning that it possesses elevated levels of hydrogen ions (low pH). It can have harmful effects on plants, aquatic animals and infrastructure. Acid rain is caused by emissions of sulfur dioxide and nitrogen oxide, which react with the water molecules in the atmosphere to produce acids. Some governments have made efforts since the 1970s to reduce the release of sulfur dioxide and nitrogen oxide into the atmosphere with positive results. Nitrogen oxides can also be produced naturally by lightning strikes, and sulfur dioxide is produced by volcanic eruptions. Acid rain has been shown to have adverse impacts on forests, freshwaters and soils, killing insect and aquatic life-forms, causing paint to peel, corrosion of steel structures such as bridges, and weathering of stone buildings and statues as well as having impacts on human health.

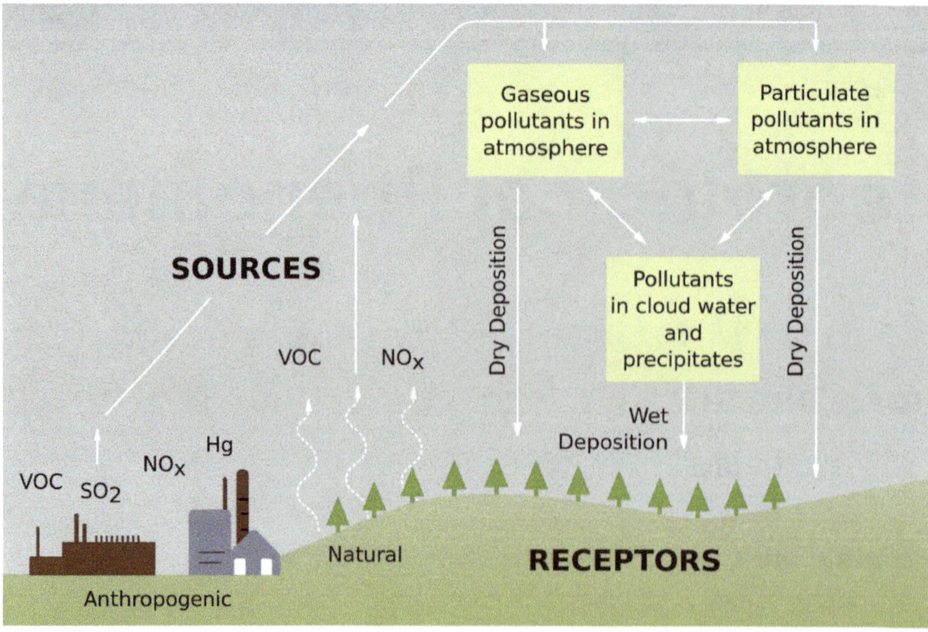

Processes involved in acid deposition (note that only SO_2 and NO_x play a significant role in acid rain).

Acid clouds can grow on SO_2 emissions from refineries, as seen here in Curaçao.

Definition

"Acid rain" is a popular term referring to the deposition of a mixture from wet (rain, snow, sleet, fog, cloudwater, and dew) and dry (acidifying particles and gases) acidic components. Distilled water, once carbon dioxide is removed, has a neutral pH of 7. Liquids with a pH less than 7 are acidic, and those with a pH greater than 7 are alkaline. "Clean" or unpolluted rain has an acidic pH, but usually no lower than 5.7, because carbon dioxide and water in the air react together to form carbonic acid, a weak acid according to the following reaction:

$$H_2O\ (l) + CO_2\ (g) \rightleftharpoons H_2CO_3\ (aq)$$

Carbonic acid then can ionize in water forming low concentrations of hydronium and carbonate ions:

$$H_2O\ (l) + H_2CO_3\ (aq) \rightleftharpoons HCO_3^-\ (aq) + H_3O^+\ (aq)$$

However, unpolluted rain can also contain other chemicals which affect its pH (acidity level). A common example is nitric acid produced by electric discharge in the atmosphere such as lightning. Acid deposition as an environmental issue (discussed later in the article) would include additional acids other than H_2CO_3.

The corrosive effect of polluted, acidic city air on limestone and marble was noted in the 17th century by John Evelyn, who remarked upon the poor condition of the Arundel marbles.Since the Industrial Revolution, emissions of sulfur dioxide and nitrogen oxides into the atmosphere have increased. In 1852, Robert Angus Smith was the first to show the relationship between acid rain and atmospheric pollution in Manchester, England.

Though acidic rain was discovered in 1853, it was not until the late 1960s that scientists began widely observing and studying the phenomenon. The term "acid rain" was coined in 1872 by Robert Angus Smith. Canadian Harold Harvey was among the first to research a "dead" lake. Public awareness of acid rain in the U.S increased in the 1970s after The New York Times published reports from the Hubbard Brook Experimental Forest in New Hampshire of the myriad deleterious environmental effects shown to result from it.

Occasional pH readings in rain and fog water of well below 2.4 have been reported in industrialized areas. Industrial acid rain is a substantial problem in China and Russia and areas downwind from them. These areas all burn sulfur-containing coal to generate heat and electricity.

The problem of acid rain has not only increased with population and industrial growth, but has become more widespread. The use of tall smokestacks to reduce local pollution has contributed to the spread of acid rain by releasing gases into regional atmospheric circulation. Often deposition occurs a considerable distance downwind of the emissions, with mountainous regions tending to receive the greatest deposition (simply because of their higher rainfall). An example of this effect is the low pH of rain which falls in Scandinavia.

History of Acid Rain in the United States

Since 1998, Harvard University wraps some of the bronze and marble statues on its campus, such as this "Chinese stele", with waterproof covers every winter, in order to protect them from erosion caused by acid rain and acid snow

The earliest report about acid rain in the United States was from the chemical evidence from Hubbard Brook Valley. In 1972, a group of scientists including Gene Likens discovered the rain that was deposited at White Mountains of New Hampshire was acidic. The pH of the sample was measured to be 4.03 at Hubbard Brook. The Hubbard Brook Ecosystem Study followed up with a series of research that analyzed the environmental effects of acid rain. Acid rain that mixed with stream water at Hubbard Brook was neutralized by the alumina from soils. The result of this research indicates the chemical reaction between acid rain and aluminum leads to increasing rate of soil weathering. Experimental research was done to examine the effects of increased acidity in stream on ecological species. In 1980, a group of scientists modified the acidity of Norris Brook, New Hampshire, and observed the change in species' behaviors. There was a decrease in species diversity, an increase in community dominants, and a decrease in the food web complexity.

In 1980, the U.S. Congress passed an Acid Deposition Act. This Act established an 18-year assessment and research program under the direction of the National Acidic Precipitation Assessment Program (NAPAP). NAPAP looked at the entire problem from a scientific perspective. It enlarged a network of monitoring sites to determine how acidic the precipitation actually was, and to determine long-term trends, and established a network for dry deposition. It looked at the effects of acid rain and funded research on the effects of acid precipitation on freshwater and terrestrial ecosystems, historical buildings, monuments, and building materials. It also funded extensive studies on atmospheric processes and potential control programs.

From the start, policy advocates from all sides attempted to influence NAPAP activities to support their particular policy advocacy efforts, or to disparage those of their opponents. For the U.S. Government's scientific enterprise, a significant impact of NAPAP were lessons learned in the assessment process and in environmental research management to a relatively large group of scientists, program managers and the public.

In 1981, the National Academy of Sciences was looking into research about the controversial issues regarding acid rain. President Ronald Reagan did not place a huge attention on the issues of acid rain until his personal visit to Canada and confirmed that Canadian border suffered from the drifting pollution from smokestacks in Midwest of US. Reagan honored the agreement to Canadian Prime Minister Trudeau's enforcement of anti-pollution regulation. In 1982, US President Ronald Reagan commissioned William Nierenberg to sit on National Science Board. Nierenberg selected scientists including Gene Likens to sit on a panel to draft a report on acid rain. In 1983, the panel of scientists came up with a draft report, which concluded that acid rain is a real problem and solutions should be sought. White House Office of Science and Technology Policy reviewed the draft report and sent Fred Singer's suggestions of the report, which cast doubt on the cause of acid rain.The panelists revealed rejections against Singer's positions and submitted the report to Nirenberg in April. In May 1983, the House of Representatives voted against legislations that aimed to control sulfur emissions. There was a debate about whether Nierenberg delayed to release the report. Nierenberg himself denied the saying about his suppression of the report and explained that the withheld of the report after the House's vote was due to the fact that the report was not ready to be published.

In 1991, DENR provided its first assessment of acid rain in the United States. It reported that 5% of New England Lakes were acidic, with sulfates being the most common problem. They noted that 2% of the lakes could no longer support Brook Trout, and 6% of the lakes were unsuitable for the survival of many species of minnow. Subsequent Reports to Congress have documented chemical changes in soil and freshwater ecosystems, nitrogen saturation, decreases in amounts of nutrients in soil, episodic acidification, regional haze, and damage to historical monuments.

Meanwhile, in 1989, the U.S. Congress passed a series of amendments to the Clean Air Act. Title IV of these amendments established the Acid Rain Program, a cap and trade system designed to control emissions of sulfur dioxide and nitrogen oxides. Title IV called for a total reduction of about 10 million tons of SO_2 emissions from power plants. It was implemented in two phases. Phase I began in 1995, and limited sulfur dioxide emissions from 110 of the largest power plants to a combined total of 8.7 million tons of sulfur dioxide. One power plant in New England (Merrimack) was in Phase I. Four other plants (Newington, Mount Tom, Brayton Point, and Salem Harbor) were added under other provisions of the program. Phase II began in 2000, and affects most of the power plants in the country.

During the 1990s, research continued. On March 10, 2005, EPA issued the Clean Air Interstate Rule (CAIR). This rule provides states with a solution to the problem of power plant pollution that drifts from one state to another. CAIR will permanently cap emissions of SO_2 and NO_x in the eastern United States. When fully implemented, CAIR will reduce SO_2emissions in 28 eastern states and the District of Columbia by over 70% and NO_x emissions by over 60% from 2003 levels.

Overall, the program›s cap and trade program has been successful in achieving its goals. Since the 1990s, SO_2 emissions have dropped 40%, and according to the Pacific Research Institute, acid rain levels have dropped 65% since 1976. Conventional regulation was used in the European Union, which saw a decrease of over 70% in SO_2 emissions during the same time period.

In 2007, total SO_2 emissions were 8.9 million tons, achieving the program's long-term goal ahead of the 2010 statutory deadline.

In 2007 the EPA estimated that by 2010, the overall costs of complying with the program for businesses and consumers would be $1 billion to $2 billion a year, only one fourth of what was originally predicted. Forbes says: *In 2010, by which time the cap and trade system had been augmented by the George W. Bush administration's Clean Air Interstate Rule, SO2 emissions had fallen to 5.1 million tons.*

The term Citizen science can be traced back as far as January 1989 and a campaign by the Audubon Society to measure Acid rain. Scientist Muki Haklay cites in a policy report for the Wilson Center entitled ‹Citizen Science and Policy: A European Perspective› a first use of the term ‹citizen science› by R. Kerson in the magazine MIT Technology Review from January 1989. Quoting from the Wilson Center report: The new form of engagement in science received the name citizen science. The first recorded example of the use of the term is from 1989, describing how 225 volunteers across the US collected rain samples to assist the Audubon Society in an acid-rain awareness raising campaign. The volunteers collected samples, checked for acidity, and reported back to the organization. The information was then used to demonstrate the full extent of the phenomenon.

Emissions of Chemicals Leading to Acidification

The most important gas which leads to acidification is sulfur dioxide. Emissions of nitrogen oxides which are oxidized to form nitric acid are of increasing importance due to stricter controls on emissions of sulfur containing compounds. 70 Tg(S) per year in the form of SO_2 comes from fossil fuel combustion and industry, 2.8 Tg(S) from wildfires and 7–8 Tg(S) per year from volcanoes.

Natural Phenomena

The principal natural phenomena that contribute acid-producing gases to the atmosphere are emissions from volcanoes. Thus, for example, fumaroles from the Laguna Caliente crater of Poás Volcano create extremely high amounts of acid rain and fog, with acidity as high as a pH of 2, clearing an area of any vegetation and frequently causing irritation to the eyes and lungs of inhabitants in nearby settlements. Acid-producing gasses are also created by biological processes that occur on the land, in wetlands, and in the oceans. The major biological source of sulfur containing compounds is dimethyl sulfide.

Nitric acid in rainwater is an important source of fixed nitrogen for plant life, and is also produced by electrical activity in the atmosphere such as lightning.

Acidic deposits have been detected in glacial ice thousands of years old in remote parts of the globe.

Soils of coniferous forests are naturally very acidic due to the shedding of needles, and the results of this phenomenon should not be confused with acid rain.

Human Activity

The coal-fired Gavin Power Plant in Cheshire, Ohio

The principal cause of acid rain is sulfur and nitrogen compounds from human sources, such as electricity generation, factories, and motor vehicles. Electrical power generation using coal is among the greatest contributors to gaseous pollutions that are responsible for acidic rain. The gases can be carried hundreds of kilometers in the atmosphere before they are converted to acids and deposited. In the past, factories had short funnels to let out smoke but this caused many problems locally; thus, factories now have taller smoke funnels. However, dispersal from these taller stacks causes pollutants to be carried farther, causing widespread ecological damage.

Chemical Processes

Combustion of fuels produces sulfur dioxide and nitric oxides. They are converted into sulfuric acid and nitric acid.

Gas Phase Chemistry

In the gas phase sulfur dioxide is oxidized by reaction with the hydroxyl radical via an intermolecular reaction:

$SO_2 + OH\cdot \rightarrow HOSO_2\cdot$

which is followed by:

$HOSO_2\cdot + O_2 \rightarrow HO_2\cdot + SO_3$

In the presence of water, sulfur trioxide (SO_3) is converted rapidly to sulfuric acid:

$SO_3\ (g) + H_2O\ (l) \rightarrow H_2SO_4\ (aq)$

Nitrogen dioxide reacts with OH to form nitric acid:

$NO_2 + OH\cdot \rightarrow HNO_3$

Chemistry in Cloud Droplets

When clouds are present, the loss rate of SO_2 is faster than can be explained by gas phase chemistry alone. This is due to reactions in the liquid water droplets.

Hydrolysis

Sulfur dioxide dissolves in water and then, like carbon dioxide, hydrolyses in a series of equilibrium reactions:

$SO_2\ (g) + H_2O \rightleftharpoons SO_2{\cdot}H_2O$

$SO_2{\cdot}H_2O \rightleftharpoons H^+ + HSO_3^-$

$HSO_3^- \rightleftharpoons H^+ + SO_3^{2-}$

Oxidation

There are a large number of aqueous reactions that oxidize sulfur from S(IV) to S(VI), leading to the formation of sulfuric acid. The most important oxidation reactions are with ozone, hydrogen peroxide and oxygen (reactions with oxygen are catalyzed by iron and manganese in the cloud droplets).

Acid Deposition

Wet Deposition

Wet deposition of acids occurs when any form of precipitation (rain, snow, and so on.) removes acids from the atmosphere and delivers it to the Earth's surface. This can result from the deposition of acids produced in the raindrops (see aqueous phase chemistry above) or by the precipitation removing the acids either in clouds or below clouds. Wet removal of both gases and aerosols are both of importance for wet deposition.

Dry Deposition

Acid deposition also occurs via dry deposition in the absence of precipitation. This can be responsible for as much as 20 to 60% of total acid deposition. This occurs when particles and gases stick to the ground, plants or other surfaces.

Adverse Effects

Acid rain has been shown to have adverse impacts on forests, freshwaters and soils, killing insect and aquatic life-forms as well as causing damage to buildings and having impacts on human health.

Surface Waters and Aquatic Animals

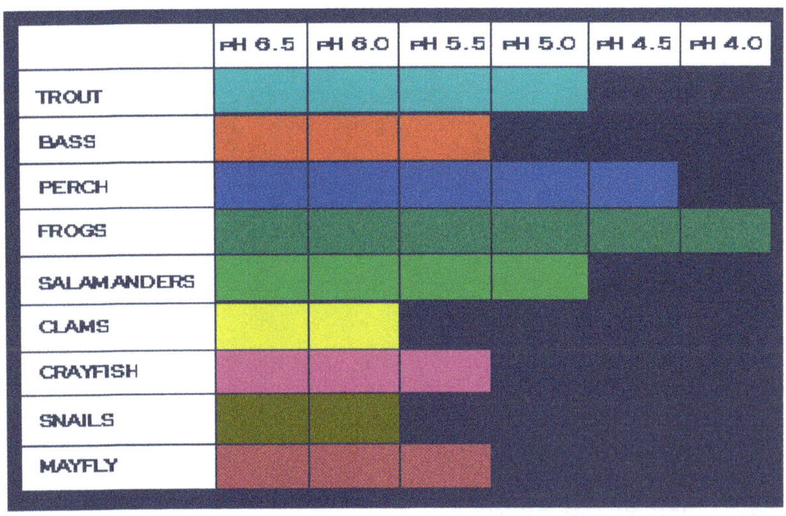

This chart shows that not all fish, shellfish, or the insects that they eat can tolerate the same amount of acid; for example, frogs can tolerate water that is more acidic (i.e., has a lower pH) than trout.

Both the lower pH and higher aluminium concentrations in surface water that occur as a result of acid rain can cause damage to fish and other aquatic animals. At pHs lower than 5 most fish eggs will not hatch and lower pHs can kill adult fish. As lakes and rivers become more acidic biodiversity is reduced. Acid rain has eliminated insect life and some fish species, including the brook trout in some lakes, streams, and creeks in geographically sensitive areas, such as the Adirondack Mountains of the United States. However, the extent to which acid rain contributes directly or indirectly via runoff from the catchment to lake and river acidity (i.e., depending on characteristics of the surrounding watershed) is variable. The United States Environmental Protection Agency's (EPA) website states: "Of the lakes and streams surveyed, acid rain caused acidity in 75% of the acidic lakes and about 50% of the acidic streams".

Soils

Soil biology and chemistry can be seriously damaged by acid rain. Some microbes are unable to tolerate changes to low pH and are killed. The enzymes of these microbes are denatured (changed in shape so they no longer function) by the acid. The hydronium ions of acid rain also mobilize toxins such as aluminium, and leach away essential nutrients and minerals such as magnesium.

$$2\ H^+\ (aq) + Mg^{2+}\ (clay) \rightleftharpoons 2\ H^+\ (clay) + Mg^{2+}\ (aq)$$

Soil chemistry can be dramatically changed when base cations, such as calcium and magnesium, are leached by acid rain thereby affecting sensitive species, such as sugar maple (Acer saccharum).

Forests and Other Vegetation

Acid rain can have severe effects on vegetation. A forest in the Black Triangle in Europe.

Adverse effects may be indirectly related to acid rain, like the acid's effects on soil (see above) or high concentration of gaseous precursors to acid rain. High altitude forests are especially vulnerable as they are often surrounded by clouds and fog which are more acidic than rain.

Other plants can also be damaged by acid rain, but the effect on food crops is minimized by the application of lime and fertilizers to replace lost nutrients. In cultivated areas, limestone may also be added to increase the ability of the soil to keep the pH stable, but this tactic is largely unusable in the case of wilderness lands. When calcium is leached from the needles of red spruce, these trees become less cold tolerant and exhibit winter injury and even death.

Ocean Acidification

Coral's limestone skeletal is sensitive to pH drop, because the calcium carbonate, core component of the limestone dissolves in acidic (low pH) solutions.

Human Health Effects

Acid rain does not directly affect human health. The acid in the rainwater is too dilute to have direct adverse effects. However, the particulates responsible for acid rain (sulfur dioxide and nitrogen oxides) do have an adverse effect. Increased amounts of fine particulate matter in the air do contribute to heart and lung problems including asthma and bronchitis.

Other Adverse Effects

Effect of acid rain on statues

Acid rain and weathering

Acid rain can damage buildings, historic monuments, and statues, especially those made of rocks, such as limestone and marble, that contain large amounts of calcium carbonate. Acids in the rain react with the calcium compounds in the stones to create gypsum, which then flakes off.

$$CaCO_3\ (s) + H_2SO_4\ (aq) \rightleftharpoons CaSO_4\ (s) + CO_2\ (g) + H_2O\ (l)$$

The effects of this are commonly seen on old gravestones, where acid rain can cause the inscriptions to become completely illegible. Acid rain also increases the corrosion rate of metals, in particular iron, steel, copper and bronze.

Affected Areas

Places significantly impacted by acid rain around the globe include most of eastern Europe from Poland northward into Scandinavia, the eastern third of the United States, and southeastern Canada. Other affected areas include the southeastern coast of China and Taiwan.

Prevention Methods

Technical Solutions

Many coal-firing power stations use flue-gas desulfurization (FGD) to remove sulfur-containing gases from their stack gases. For a typical coal-fired power station, FGD will remove 95% or more of the SO_2 in the flue gases. An example of FGD is the wet scrubber which is commonly used. A wet scrubber is basically a reaction tower equipped with a fan that extracts hot smoke stack gases from a power plant into the tower. Lime or limestone in slurry form is also injected into the tower to mix with the stack gases and combine with the sulfur dioxide present. The calcium carbonate of the limestone produces pH-neutral calcium sulfate that is physically removed from the scrubber. That is, the scrubber turns sulfur pollution into industrial sulfates.

In some areas the sulfates are sold to chemical companies as gypsum when the purity of calcium sulfate is high. In others, they are placed in landfill. However, the effects of acid rain can last for generations, as the effects of pH level change can stimulate the continued leaching of undesirable chemicals into otherwise pristine water sources, killing off vulnerable insect and fish species and blocking efforts to restore native life.

Fluidized bed combustion also reduces the amount of sulfur emitted by power production.

Vehicle emissions control reduces emissions of nitrogen oxides from motor vehicles.

International Treaties

A number of international treaties on the long-range transport of atmospheric pollutants have been agreed for example, the 1985 Helsinki Protocol on the Reduction of Sulphur Emissions under the Convention on Long-Range Transboundary Air Pollution. Canada and the US signed the Air Quality Agreement in 1991. Most European countries and Canada have signed the treaties.

Emissions Trading

In this regulatory scheme, every current polluting facility is given or may purchase on an open market an emissions allowance for each unit of a designated pollutant it emits. Operators can then install pollution control equipment, and sell portions of their emissions allowances they no longer need for their own operations, thereby recovering some of the capital cost of their investment in such equipment. The intention is to give operators economic incentives to install pollution controls.

The first emissions trading market was established in the United States by enactment of the Clean Air Act Amendments of 1990. The overall goal of the Acid Rain Program established by the Act is to achieve significant environmental and public health benefits through reductions in emissions of sulfur dioxide (SO_2) and nitrogen oxides (NO_x), the primary causes of acid rain. To achieve this goal at the lowest cost to society, the program employs both regulatory and market based approaches for controlling air pollution.

AIR MASS

In meteorology, an air mass is a volume of air defined by its temperature and water vapor content. Air masses cover many hundreds or thousands of square miles, and adapt to the characteristics of the surface below them. They are classified according to latitude and their continental or maritime source regions. Colder air masses are termed polar or arctic, while warmer air masses are deemed tropical. Continental and superior air masses are dry while maritime and monsoon air masses are moist. Weather fronts separate air masses with different density (temperature and/or moisture) characteristics. Once an air mass moves away from its source region, underlying vegetation and water bodies can quickly modify its character. Classification schemes tackle an air mass' characteristics, and well as modification.

Classification and Notation

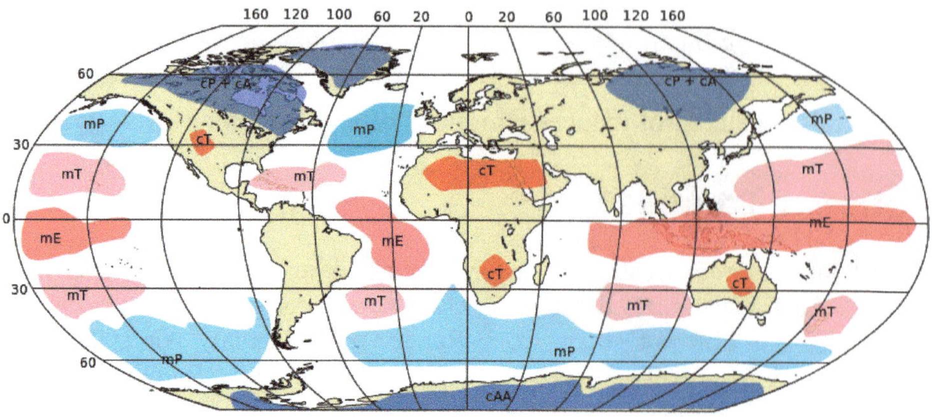

Source regions of global air masses

The Bergeron classification is the most widely accepted form of air mass classification, though others have produced more refined versions of this scheme over different regions of the globe. Air mass classification involves three letters. The first letter describes its moisture properties, with c used for continental air masses (dry) and m for maritime air masses (moist). The second letter describes the thermal characteristic of its source region: T for Tropical, P for Polar, A for arctic or Antarctic, M for monsoon, E for Equatorial, and S for superior air (an adiabatically drying and warming air formed by significant downward motion in the atmosphere). For instance, an air mass originating over the desert southwest of the United States in summer may be designated “cT”. An air mass originating over northern Siberia in winter may be indicated as “cA”.

The stability of an air mass may be shown using a third letter, either “k” (air mass colder than the surface below it) or “w” (air mass warmer than the surface below it). An example of this might be a polar air mass blowing over the Gulf Stream, denor it is replacing is colder than the replaced air mass (usually for polar air masses). For example, a series of fronts over the Pacific might show an air mass denoted mPk followed by another denoted mPk’.

Another convention utilizing these symbols is the indication of modification or transformation of one type to another. For instance, an Arctic air mass blowing out over the Gulf of Alaska may be shown as “cA-mPk”. Yet another convention indicates the layering of air masses in certain situations. For instance, the overrunning of a polar air mass by an air mass from the Gulf of Mexico over the Central United States might be shown with the notation “mT/cP” (sometimes using a horizontal line as in fraction notation).

Characteristics

Arctic, Antarctic, and polar air masses are cold. The qualities of arctic air are developed over ice and snow-covered ground. Arctic air is deeply cold, colder than polar air masses. Arctic air can be shallow in the summer, and rapidly modify as it moves equatorward. Polar air masses develop over higher latitudes over the land or ocean, are very stable, and generally shallower than arctic air. Polar air over the ocean (maritime) loses its stability as it gains moisture over warmer ocean waters.

Tropical and equatorial air masses are hot as they develop over lower latitudes. Those that develop over land (continental) are drier and hotter than those that develop over oceans, and travel poleward on the western periphery of the subtropical ridge. Maritime tropical air masses are sometimes referred to as trade air masses. Monsoon air masses are moist and unstable. Superior air masses are dry, and rarely reach the ground. They normally reside over maritime tropical air masses, forming a warmer and drier layer over the more moderate moist air mass below, forming what is known as a trade wind inversion over the maritime tropical air mass. Continental Polar air masses (cP) are air masses that are cold and dry due to their continental source region. Continental polar air masses that affect North America form over interior Canada. Continental Tropical air masses (cT) are a type of tropical air produced by the subtropical ridge over large areas of land and typically originate from low-latitude deserts such as the Sahara Desert in northern Africa, which is the major source of these air masses. Other less important sources producing cT air masses are the Arabian Peninsula, the central arid/semi-arid part of Australia and deserts lying in the Southwestern United States. Continental tropical air masses are extremely hot and dry.

Movement and Fronts

A **weather front** is a boundary separating two masses of air of different densities, and is the principal cause of meteorological phenomena. In surface weather analyses, fronts are depicted using various colored lines and symbols, depending on the type of front. The air masses separated by a front usually differ in temperature and humidity. Cold fronts may feature narrow bands of thunderstorms and severe weather, and may on occasion be preceded by squall lines or dry lines. Warm fronts are usually preceded by stratiform precipitation and fog. The weather usually clears quickly after a front's passage. Some fronts produce no precipitation and little cloudiness, although there is invariably a wind shift.

Cold fronts and occluded fronts generally move from west to east, while warm fronts move poleward. Because of the greater density of air in their wake, cold fronts and cold occlusions move faster than warm fronts and warm occlusions. Mountains and warm bodies of water can slow the movement of fronts. When a front becomes stationary, and the density contrast across the frontal boundary vanishes, the front can degenerate into a line which separates regions of differing wind velocity, known as a shearline. This is most common over the open ocean.

Modification

Lake-effect snow bands near the Korean Peninsula

Air masses can be modified in a variety of ways. Surface flux from underlying vegetation, such as forest, acts to moisten the overlying air mass.Heat from underlying warmer waters can significantly modify an air mass over distances as short as 35 kilometres (22 mi) to 40 kilometres (25 mi). For example, southwest of extratropical cyclones, curved cyclonic flow bringing cold air across the relatively warm water bodies can lead to narrow lake-effect snow bands. Those bands bring strong localized precipitation since large water bodies such as lakes efficiently store heat that results in significant temperature differences (larger than 13 °C or 23 °F) between the water surface and the air above. Because of this temperature difference, warmth and moisture are transported upward, condensing into vertically oriented clouds (see satellite picture) which produce snow showers. The temperature decrease with height and cloud depth are directly affected by both the water temperature and the large-scale environment. The stronger the temperature decrease with height, the deeper the clouds get, and the greater the precipitation rate becomes.

POLAR VORTEX

A polar vortex is an upper level low-pressure area lying near the Earth's pole. There are two polar vortices in the Earth's atmosphere, which overlie the North, and South Poles. Each polar vortex is a persistent, large-scale, low pressure zone that rotates counter-clockwise at the North Pole (called a cyclone), and clockwise at the South Pole. The bases of the two polar vortices are located in the middle and upper troposphere and extend into the stratosphere. Beneath that lies a large mass of cold, dense arctic air. The vortices weaken and strengthen from year to year. When the vortex of the arctic is strong it is well defined, there is a single vortex and the arctic air is well contained; when weaker, which it generally is, it will break into two or more vortices; when very weak, the flow of arctic air becomes more disorganized and masses of cold arctic air can push equatorward, bringing with it a rapid and sharp temperature drop. The interface between the cold dry air mass of the pole and the warm moist air mass further south defines the location of the polar front. The polar front is centered, roughly at 60° latitude. A polar vortex strengthens in the winter and weakens in the summer due to its dependence on the temperature difference between the equator and the poles. The vortices span less than 1,000 kilometers (620 miles) in diameter within which they rotate counter-clockwise in the Northern Hemisphere, and in a clockwise fashion in the Southern Hemisphere. As with other cyclones, their rotation is driven by the Coriolis effect.

When the polar vortex is strong, there is a single vortex with a jet stream that is "well constrained" near the polar front. When the northern vortex weakens, it separates into two or more vortices, the strongest of which are near Baffin Island, Canada and the other over northeast Siberia. The Antarctic vortex of the Southern Hemisphere is a single low pressure zone that is found near the edge of the Ross ice shelf near 160 west longitude. When the polar vortex is strong, the mid-latitude Westerlies (winds at the surface level between 30° and 60° latitude from the west) increase in strength and are persistent. When the polar vortex is weak, high pressure zones of the mid latitudes may push poleward, moving the polar vortex, jet stream, and polar front equatorward. The jet stream is seen to "buckle" and deviate south. This rapidly brings cold dry air into contact with the warm, moist air of the mid latitudes, resulting in a rapid and dramatic change of weather known as a "cold snap".

Ozone depletion occurs within the polar vortices – particularly over the Southern Hemisphere – reaching a maximum depletion in the spring.

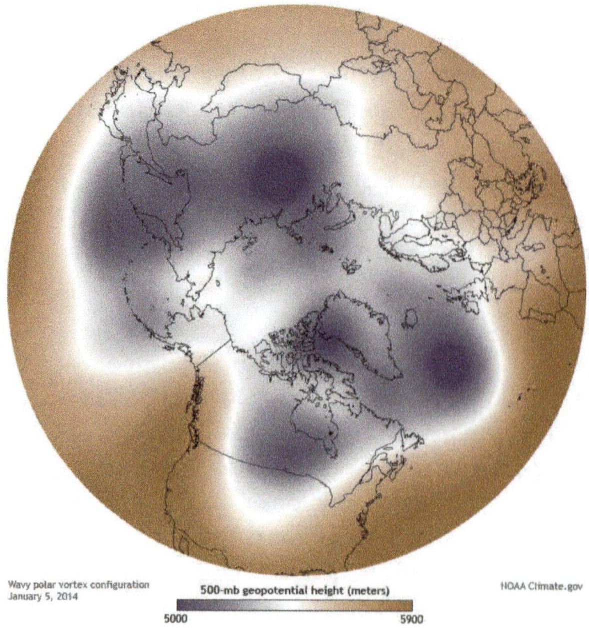

The typical polar vortex configuration in November, 2013...

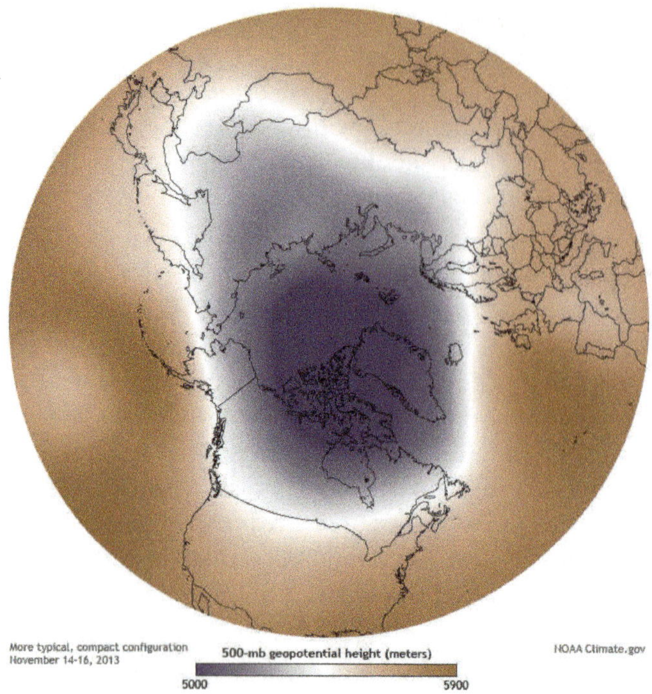

A weak polar vortex on January 5, 2014.

Identification

Polar cyclones are low pressure zones embedded within the polar air masses, and exist year-round. The stratospheric polar vortex develops at latitudes above the subtropical jet stream.Horizontally, most polar vortices have a radius of less than 1,000 kilometres (620 mi). Since polar vortices exist from the stratosphere downward into the mid-troposphere, a variety of heights/pressure levels are used to mark its position. The 50 mb pressure surface is most often used to identify its stratospheric location. At the level of the tropopause, the extent of closed contours of potential temperature can be used to determine its strength. Others have used levels down to the 500 hPa pressure level (about 5,460 metres (17,910 ft) above sea level during the winter) to identify the polar vortex.

Duration and Power

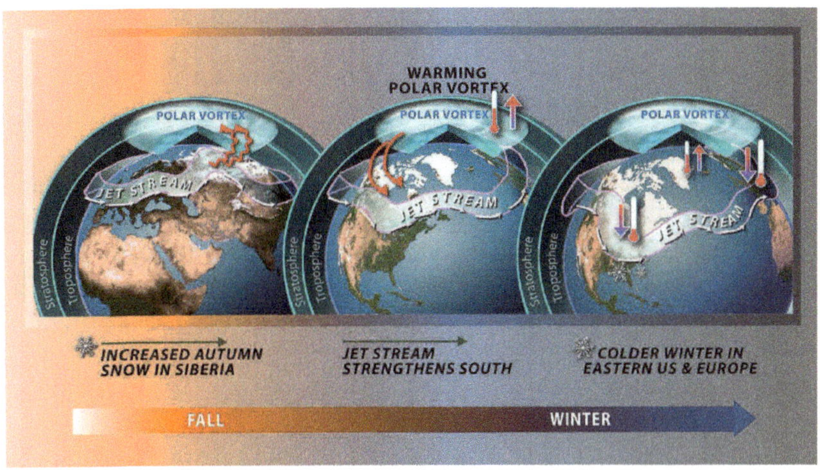

Polar vortex and weather impacts due to stratospheric warming

Polar vortices are weakest during summer and strongest during winter. Extratropical cyclones that migrate into higher latitudes when the polar vortex is weak can disrupt the single vortex creating smaller vortices (cold-core lows) within the polar air mass. Those individual vortices can persist for more than a month.

Volcanic eruptions in the tropics can lead to a stronger polar vortex during winter for as long as two years afterwards. The strength and position of the polar vortex shapes the flow pattern in a broad area about it. An index which is used in the northern hemisphere to gauge its magnitude is the Arctic oscillation.

When the Arctic vortex is at it strongest, there is a single vortex, but normally, the Arctic vortex is elongated in shape, with two cyclone centers, one over Baffin Island in Canada and the other over northeast Siberia. When the Arctic pattern is at its weakest, subtropic air masses can intrude poleward causing the Arctic air masses to move equatorward, as during the Winter 1985 Arctic outbreak. The Antarctic polar vortex is more pronounced and persistent than the Arctic one. In the Arctic the distribution of land masses at high latitudes in the Northern Hemisphere gives rise to Rossby waves which contribute to the breakdown of the polar vortex, whereas in the Southern Hemisphere the vortex is less disturbed. The breakdown of the polar vortex is an extreme event known as a sudden stratospheric warming, here the vortex completely breaks down and an associated warming of 30–50 °C (54–90 °F) over a few days can occur.

The waxing and waning of the polar vortex is driven by the movement of mass and the transfer of heat in the polar region. In the autumn, the circumpolar winds increase in speed and the polar vortex rises into the stratosphere. The result is the polar air forms a coherent rotating air mass: the polar vortex. As winter approaches, the vortex core cools, the winds decrease, and the vortex energy declines. Once late winter and early spring approach the vortex is at its weakest. As a result, during late winter, large fragments of the vortex air can be diverted into lower latitudes by stronger weather systems intruding from those latitudes. In the lowest level of the stratosphere, strong potential vorticity gradients remain, and the majority of that air remains confined within the polar air mass into December in the Southern Hemisphere and April in the Northern Hemisphere, well after the breakup of the vortex in the mid-stratosphere.

The breakup of the northern polar vortex occurs between mid March to mid May. This event signifies the transition from winter to spring, and has impacts on the hydrological cycle, growing seasons of vegetation, and overall ecosystem productivity. The timing of the transition also influences changes in sea ice, ozone, air temperature, and cloudiness. Early and late polar breakup episodes have occurred, due to variations in the stratospheric flow structure and upward spreading of planetary waves from the troposphere. As a result of increased waves into the vortex, the vortex experiences more rapid warming than normal, resulting in an earlier breakup and spring. When the breakup comes early, it is characterized by with persistent of remnants of the vortex. When the breakup is late, the remnants dissipate rapidly. When the breakup is early, there is one warming period from late February to middle March. When the breakup is late, there are two warming periods, one January, and one in March. Zonal mean temperature, wind, and geopotential height exert varying deviations from their normal values before and after early breakups, while the deviations remain constant before and after late breakups. Scientists are connecting a delay in the Arctic vortex breakup with a reduction of planetary wave activities, few stratospheric sudden warming events, and depletion of ozone.

Sudden stratospheric warming events are associated with weaker polar vortices. This warming of stratospheric air can reverse the circulation in the Arctic Polar Vortex from counter-clockwise to clockwise. These changes aloft force changes in the troposphere below. An example of an effect on the troposphere is the change in speed of the Atlantic Ocean circulation pattern. A soft spot just south of Greenland is where the initial step of downwelling occurs, nicknamed the "Achilles Heel of the North Atlantic". Small amounts of heating or cooling traveling from the polar vortex can trigger or delay downwelling, altering the Gulf Stream Current of the Atlantic, and the speed of other ocean currents. Since all other oceans depend on the Atlantic Ocean›s movement of heat energy, climates across the planet can be dramatically affected. The weakening or strengthening of the polar vortex can alter the sea circulation more than a mile beneath the waves. Strengthening storm systems within the troposphere that cool the poles, intensify the polar vortex. La Niña–related climate anomalies significantly strengthen the polar vortex. Intensification of the polar vortex produces changes in relative humidity as downward intrusions of dry, stratospheric air enter the vortex core. With a strengthening of the vortex comes a longwave cooling due to a decrease in water vapor concentration near the vortex. The decreased water content is a result of a lower tropopause within the vortex, which places dry stratospheric air above moist tropospheric air. Instability is caused when the vortex tube, the line of concentrated vorticity, is displaced. When this occurs, the vortex rings become more unstable and prone to shifting by planetary waves.The planetary wave activity in both hemispheres varies year-to-year, producing a corresponding response in the strength and temperature of the polar vortex. The number of waves around the perimeter of the vortex are related to the core size; as the vortex core decreases, the number of waves increase.

The degree of the mixing of polar and mid-latitude air depends on the evolution and position of the polar night jet. In general, the mixing is less inside the vortex than outside. Mixing occurs with unstable planetary waves that are characteristic of the middle and upper stratosphere in winter. Prior to vortex breakdown, there is little transport of air out of the Arctic Polar Vortex due to strong barriers above 420 km (261 miles). The polar night jet which exists below this, is weak in the early winter. As a result, it does not deviate any descending polar air, which then mixes with air in the mid-latitudes. In the late winter, air parcels do not descend as much, reducing mixing. After the vortex is broken up, the ex-vortex air is dispersed into the middle latitudes within a month.

Sometimes, a mass of the polar vortex breaks off before the end of the final warming period. If large enough, the piece can move into Canada and the Midwestern, Central, Southern, and Northeastern United States. This diversion of the polar vortex can occur due to the displacement of the polar jet stream; for example, the significant northwestward direction of the polar jet stream in the western part of the United States during the winters of 2013–2014, and 2014-2015. This caused warm, dry conditions in the west, and cold, snowy conditions in the east.Occasionally, the high-pressure air mass, called the Greenland Block, can cause the polar vortex to divert to the south, rather than follow its normal path over the North Atlantic.

Climate Change

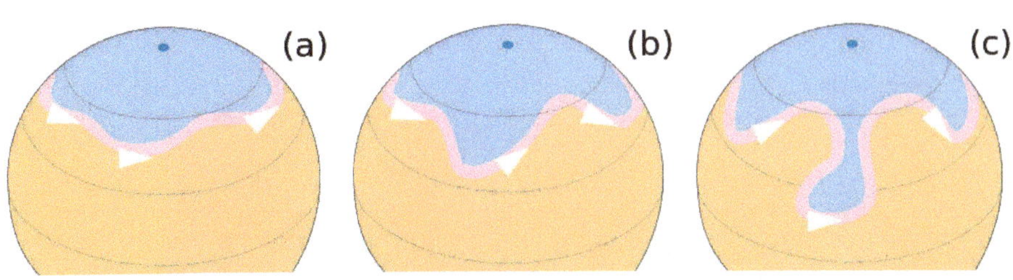

Meanders of the northern hemisphere's jet stream developing (a, b) and finally detaching a "drop" of cold air (c); orange: warmer masses of air; pink: jet stream

A study in 2001 found that stratospheric circulation can have anomalous effects on weather regimes. In the same year researchers found a statistical correlation between weak polar vortex and outbreaks of severe cold in the Northern Hemisphere. In more recent years scientists identified interactions with Arctic sea ice decline, reduced snow cover, evapotranspiration patterns, NAO anomalies or weather anomalies which are linked to the polar vortex and jet stream configuration. However, because the specific observations are considered short-term observations (starting c. 13 years ago) there is considerable uncertainty in the conclusions. Climatology observations require several decades to definitively distinguish natural variability from climate trends.

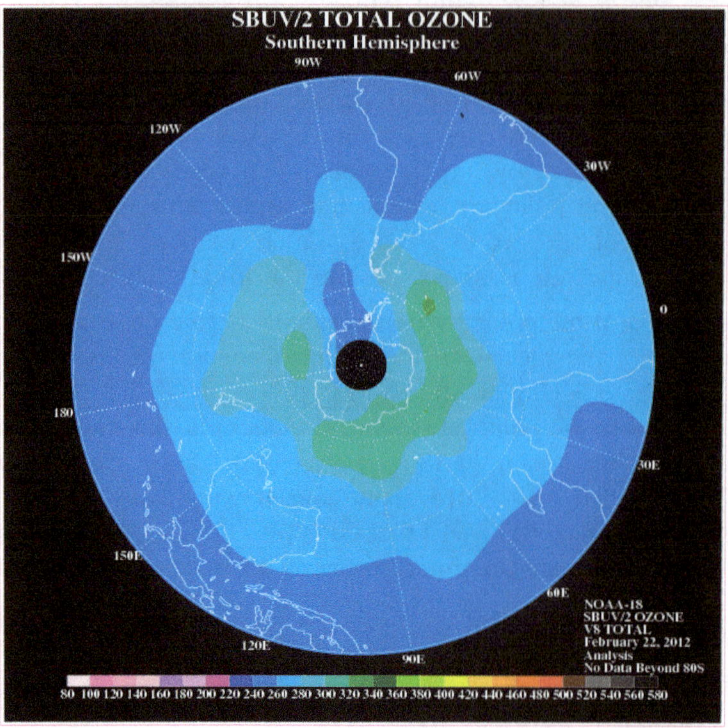

Southern Hemisphere Ozone Concentration, February 22, 2012

The general assumption is that reduced snow cover and sea ice reflect less sunlight and therefore evaporation and transpiration increases, which in turn alters the pressure and temperature gradient of the polar vortex, causing it to weaken or collapse. This becomes apparent when the jet stream amplitude increases (meanders) over the northern hemisphere, causing Rossby waves to propagate farther to the south or north, which in turn transports warmer air to the north pole and polar air into lower latitudes. The jet stream amplitude increases with a weaker polar vortex, hence increases the chance for weather systems to become blocked. A recent blocking event emerged when a high-pressure over Greenland steered Hurricane Sandy into the northern Mid-Atlantic states.

Ozone Depletion

The chemistry of the Antarctic polar vortex has created severe ozone depletion. The nitric acid in polar stratospheric clouds reacts with chlorofluorocarbons to form chlorine, which catalyzes the photochemical destruction of ozone. Chlorine concentrations build up during the polar winter, and the consequent ozone destruction is greatest when the sunlight returns in spring. These clouds can only form at temperatures below about −80 °C (−112 °F). Since there is greater air exchange between the Arctic and the mid-latitudes, ozone depletion at the north pole is much less severe than at the south. Accordingly, the seasonal reduction of ozone levels over the Arctic is usually characterized as an "ozone dent", whereas the more severe ozone depletion over the Antarctic is considered an "ozone hole". This said, chemical ozone destruction in the 2011 Arctic polar vortex attained, for the first time, a level clearly identifiable as an Arctic "ozone hole".

Outside Earth

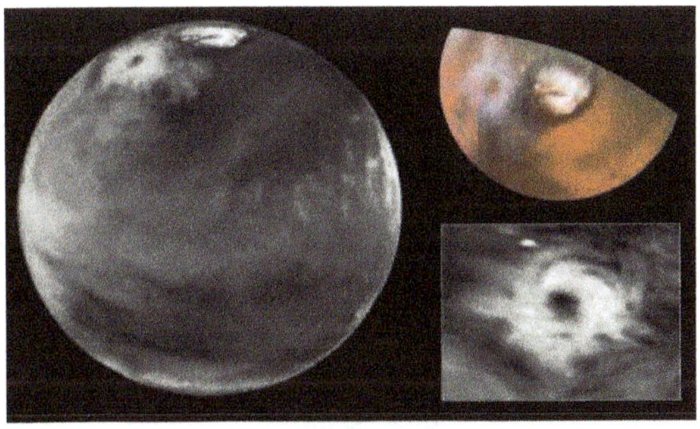

Hubble view of the colossal polar cloud on Mars

Other astronomical bodies are also known to have polar vortices, including Venus (double vortex—that is, two polar vortices at a pole), Mars, Jupiter, Saturn, and Saturn's moon Titan.

Hot Polar Vortex

Saturn's south pole is the only known hot polar vortex in the solar system.

BALL LIGHTNING

Ball lightning is an unexplained atmospheric electrical phenomenon. The term refers to reports of luminous, spherical objects that vary from pea-sized to several meters in diameter. Though usually associated with thunderstorms, the phenomenon lasts considerably longer than the split-second flash of a lightning bolt. Many early reports claim that the ball eventually explodes, sometimes with fatal consequences, leaving behind the odor of sulfur.

Until the 1960s, most scientists argued that ball lightning was not a real phenomenon but an urban myth, despite numerous reports from throughout the world. Laboratory experiments can produce effects that are visually similar to reports of ball lightning, but how these relate to the natural phenomenon remains unclear.

Scientists have proposed many hypotheses about ball lightning over the centuries. Scientific data on natural ball lightning are scarce, owing to its infrequency and unpredictability. The presumption of its existence depends on reported public sightings, and has therefore produced somewhat inconsistent findings. Owing to inconsistencies and to the lack of reliable data, the true nature of ball lightning remains unknown. The first ever optical spectrum of what appears to have been a ball-lightning event was published in January 2014 and included a video at high frame-rate.

A 1901 depiction of ball lightning

The Great Thunderstorm of Widecombe-in-the-Moor

Another early description was reported during the Great Thunderstorm at a church in Widecombe-in-the-Moor, Devon, in England, on 21 October 1638. Four people died and approximately 60 were injured when, during a severe storm, an 8-foot (2.4 m) ball of fire was described as striking and entering the church, having nearly destroyed it. Large stones from the church walls were hurled into the ground and through large wooden beams. The ball of fire allegedly smashed the pews and many windows, and filled the church with a foul sulfurous odour and dark, thick smoke.

The ball of fire reportedly divided into two segments, one exiting through a window by smashing it open, the other disappearing somewhere inside the church. The explanation at the time, because of the fire and sulfur smell, was that the ball of fire was "the devil" or the “flames of hell”. Later, some blamed the entire incident on two people who had been playing cards in the pew during the sermon, thereby incurring God’s wrath.

The *Catherine and Mary*

In December 1726 a number of British newspapers printed an extract of a letter from John Howell of the sloop *Catherine and Mary*:

As we were coming thro’ the Gulf of Florida on 29th of August, a large ball of fire fell from the Element and split our mast in Ten Thousand Pieces, if it were possible; split our Main Beam, also Three Planks of the Side, Under Water, and Three of the Deck; killed one man, another had his Hand carried of, and had it not been for the violent rains, our Sails would have been of a Blast of Fire.

The *Montague*

One particularly large example was reported “on the authority of Dr. Gregory” in 1749:

Admiral Chambers on board the Montague, 4 November 1749, was taking an observation just before noon...he observed a large ball of blue fire about three miles distant from them. They immediately lowered their topsails, but it came up so fast upon them, that, before they could raise the main tack, they observed the ball rise almost perpendicularly, and not above forty or fifty yards from the main chains when it went off with an explosion, as great as if a hundred cannons had been discharged at the same time, leaving behind it a strong sulphurous smell. By this explosion the main top-mast was shattered into pieces and the main mast went down to the keel.

Five men were knocked down and one of them very bruised. Just before the explosion, the ball seemed to be the size of a large mill-stone.

Georg Richmann

A 1753 report depicts ball lightning as having been lethal when Professor Georg Richmann of Saint Petersburg, Russia, created a kite-flying apparatus similar to Benjamin Franklin’s proposal a year earlier. Richmann was attending a meeting of the Academy of Sciences when he heard thunder and ran home with his engraver to capture the event for posterity. While the experiment was under way, ball lightning appeared and travelled down the string, struck Richmann›s forehead and killed him. The ball had left a red spot on Richmann›s forehead, his shoes were blown open, and his clothing was singed. His engraver was knocked unconscious. The door frame of the room was split and the door was torn from its hinges.

HMS *Warren Hastings*

An English journal reported that during an 1809 storm, three “balls of fire” appeared and “attacked” the British ship *HMS Warren Hastings*. The crew watched one ball descend, killing a man on deck and setting the main mast on fire. A crewman went out to retrieve the fallen body and was struck by a second ball, which knocked him back and left him with mild burns. A third man was killed by contact with the third ball. Crew members reported a persistent, sickening sulfur smell afterward.

Ebenezer Cobham Brewer

Ebenezer Cobham Brewer, in his 1864 US edition of *A Guide to the Scientific Knowledge of Things Familiar*, discussed "globular lightning". He describes it as slow-moving balls of fire or explosive gas that sometimes fall to the earth or run along the ground during a thunderstorm. He said that the balls sometimes split into smaller balls and may explode "like a cannon".

Wilfrid De Fonvielle

In his book *Thunder and Lighting*, translated into English in 1875, French science writer Wilfrid de Fonvielle wrote that there had been about 150 reports of globular lightning:

Globular lighting seems to be particularly attracted to metals; thus it will seek the railings of balconies, or else water or gas pipes etc, It has no peculiar tint of its own but will appear of any colour as the case may be ... at Coethen in the Duchy of Anhalt it appeared green. M. Colon, Vice-President of the Geological Society of Paris, saw a ball of lightning descend slowly from the sky along the bark of a poplar tree; as soon as it touched the earth it bounced up again, and disappeared without exploding. On 10th of September 1845 a ball of lightning entered the kitchen of a house in the village of Salagnac in the valley of Correze. This ball rolled across without doing any harm to two women and a young man who were here; but on getting into an adjoining stable it exploded and killed a pig which happened to be shut up there, and which, knowing nothing about the wonders of thunder and lightning, dared to smell it in the most rude and unbecoming manner.

The motion of such balls is far from being very rapid – they have even been observed occasionally to pause in their course, but they are not the less destructive for all that. A ball of lightning which entered the church of Stralsund, on exploding, projected a number of balls which exploded in their turn like shells.

Tsar Nicholas II

Tsar Nicholas II, the last Emperor of Russia, reported witnessing what he called "a fiery ball" while in the company of his grandfather, Tsar Alexander II: "Once my parents were away," recounted the Tsar, "and I was at the all-night vigil with my grandfather in the small church in Alexandria. During the service there was a powerful thunderstorm, streaks of lightning flashed one after the other, and it seemed as if the peals of thunder would shake even the church and the whole world to its foundations. Suddenly it became quite dark, a blast of wind from the open door blew out the flame of the candles which were lit in front of the iconostasis, there was a long clap of thunder, louder than before, and I suddenly saw a fiery ball flying from the window straight towards the head of the Emperor. The ball (it was of lightning) whirled around the floor, then passed the chandelier and flew out through the door into the park. My heart froze, I glanced at my grandfather

– his face was completely calm. He crossed himself just as calmly as he had when the fiery ball had flown near us, and I felt that it was unseemly and not courageous to be frightened as I was. I felt that one had only to look at what was happening and believe in the mercy of God, as he, my grandfather, did. After the ball had passed through the whole church, and suddenly gone out through the door, I again looked at my grandfather. A faint smile was on his face, and he nodded his head at me. My panic disappeared, and from that time I had no more fear of storms."

Aleister Crowley

British occultist Aleister Crowley reported witnessing what he referred to as "globular electricity" during a thunderstorm on Lake Pasquaney in New Hampshire in 1916. He was sheltered in a small cottage when he "noticed, with what I can only describe as calm amazement, that a dazzling globe of electric fire, apparently between six and twelve inches (15–30 cm) in diameter, was stationary about six inches below and to the right of my right knee. As I looked at it, it exploded with a sharp report quite impossible to confuse with the continuous turmoil of the lightning, thunder and hail, or that of the lashed water and smashed wood which was creating a pandemonium outside the cottage. I felt a very slight shock in the middle of my right hand, which was closer to the globe than any other part of my body.

R.C. Jennison

Jennison, of the Electronics Laboratory at the University of Kent, described his own observation of ball lightning:

I was seated near the front of the passenger cabin of an all-metal airliner (Eastern Airlines Flight EA 539) on a late night flight from New York to Washington. The aircraft encountered an electrical storm during which it was enveloped in a sudden bright and loud electrical discharge (0005 h EST, March 19, 1963). Some seconds after this a glowing sphere a little more than 20 cm in diameter emerged from the pilot's cabin and passed down the aisle of the aircraft approximately 50 cm from me, maintaining the same height and course for the whole distance over which it could be observed.

Other Accounts

Ball lightning entering via the chimney (1886)

- On 30 April 1877, a ball of lightning entered the Golden Temple at Amritsar, India, and exited through a side door. Several people observed the ball, and the incident is inscribed on the front wall of Darshani Deodhi.
- On 22 November 1894, an unusually prolonged instance of natural ball lightning occurred in Golden, Colorado, which suggests it could be artificially induced from the atmosphere. The *Golden Globe* newspaper reported, "A beautiful yet strange phenomenon was seen in this city on last Monday night. The wind was high and the air seemed to be full of electricity. In front of, above and around the new Hall of Engineering of the School of Mines, balls of fire played tag for half an hour, to the wonder and amazement of all who saw the display. In this building is situated the dynamos and electrical apparatus of perhaps the finest electrical plant of its size in the state. There was probably a visiting delegation from the clouds, to the captives of the dynamos on last Monday night, and they certainly had a fine visit and a roystering game of romp.
- On 22 May 1901 in the Kazakh city of Ouralsk in the Russian Empire (now Oral, Kazakhstan), "a dazzlingly brilliant ball of fire" descended gradually from the sky during a thunderstorm, then entered into a house where 21 people had taken refuge, "wreaked havoc with the apartment, broke through the wall into a stove in the adjoining room, smashed the stove-pipe, and carried it off with such violence that it was dashed against the opposite wall, and went out through the broken window". The incident was reported in the *Bulletin of the Société astronomique de France* the following year.
- In July 1907 the Cape Naturaliste Lighthouse in Western Australia was hit by ball lightning. Lighthouse keeper Patrick Baird was in the tower at the time and was knocked unconscious. His daughter Ethel recorded the event.
- An early fictional reference to ball lightning appears in a children›s book set in the 19th century by Laura Ingalls Wilder. The books are considered historical fiction, but the author always insisted they were descriptive of actual events in her life. In Wilder›s description, three separate balls of lightning appear during a winter blizzard near a cast iron stove in the family›s kitchen. They are described as appearing near the stovepipe, then rolling across the floor, only to disappear as the mother (Caroline Ingalls) chases them with a willow-branch broom.
- Pilots in World War II described an unusual phenomenon for which ball lightning has been suggested as an explanation. The pilots saw small balls of light moving in strange trajectories, which came to be referred to as foo fighters.
- Submariners in WWII gave the most frequent and consistent accounts of small ball lightning in the confined submarine atmosphere. There are repeated accounts of inadvertent production of floating explosive balls when the battery banks were switched in or out, especially if mis-switched or when the highly inductive electrical motors were mis-connected or disconnected. An attempt later to duplicate those balls with a surplus submarine battery resulted in several failures and an explosion.
- On 6 August 1944, a ball of lightning went through a closed window in Uppsala, Sweden, leaving a circular hole about 5 centimetres (2.0 in) in diameter. The

incident was witnessed by residents in the area, and was recorded by a lightning strike tracking system on the Division for Electricity and Lightning Research at Uppsala University.

- In 1954 Domokos Tar, a physicist, observed a lightning strike during a heavy thunderstorm. A single bush was flattened in the wind. Some seconds later a speedy rotating ring (cylinder) appeared in the shape of a wreath. The ring was about 5 metres (16 ft) away from the lightning impact point. The ring›s plane was perpendicular to the ground and in full view of the observer. The outer/inner diameters were about 60 and 30 centimetres (24 and 12 in) respectively. The ring rotated quickly about 80 cm (31 in) above the ground. It was composed of wet leaves and dirt and rotated counter clockwise. After seconds the ring became self-illuminated turning increasingly red, then orange, yellow and finally white. The ring (cylinder) at the outside was similar to a sparkler. In spite of the rain, many electrical high voltage discharges could be seen. After some seconds, the ring suddenly disappeared and simultaneously the ball lightning appeared in the middle. Initially the ball had only one tail and it rotated in the same direction as the ring. It was homogeneous and showed no transparency. In the first moment the ball hovered motionless, but then began to move forward on the same line with a constant speed of about 1 metre per second (3.3 ft/s). It was stable and travelled at the same height in spite of the heavy rain and strong wind. After moving about 10 m (33 ft) it suddenly disappeared without any noise.
- On 10 July 2011, during a powerful thunderstorm, a ball of light with a two-metre (6 ft 7 in) tail went through a window to the control room of local emergency services in Liberec, Czech Republic. The ball bounced from window to the ceiling, then to the floor and back to the ceiling, where it rolled along it for two or three metres. Then it dropped to the floor and disappeared. The staff present in the control room were frightened, smelled electricity and burned cables and thought something was burning. The computers froze (not crashed) and all communications equipment was knocked out for the night until restored by technicians. Aside from damages caused by disrupting equipment, only one computer monitor was destroyed.
- On Dec 15th 2014, flight BE-6780 (Saab 2000) in the UK experienced ball lightning in the forward cabin just before lightning struck the aircraft nose.

Characteristics

Descriptions of ball lightning vary widely. It has been described as moving up and down, sideways or in unpredictable trajectories, hovering and moving with or against the wind; attracted to, unaffected by, or repelled from buildings, people, cars and other objects. Some accounts describe it as moving through solid masses of wood or metal without effect, while others describe it as destructive and melting or burning those substances. Its appearance has also been linked to power lines as well as during thunderstorms and also calm weather. Ball lightning has been described as transparent, translucent, multicolored, evenly lit, radiating flames, filaments or sparks, with shapes that vary between spheres, ovals, tear-drops, rods, or disks.

Ball lightning is often erroneously identified as St. Elmo's fire. They are separate and distinct phenomena.

The balls have been reported to disperse in many different ways, such as suddenly vanishing, gradually dissipating, absorption into an object, "popping," exploding loudly, or even exploding with force, which is sometimes reported as damaging. Accounts also vary on their alleged danger to humans, from lethal to harmless.

A review of the available literature published in 1972 identified the properties of a "typical" ball lightning, whilst cautioning against over-reliance on eye-witness accounts:

- They frequently appear almost simultaneously with cloud-to-ground lightning discharge
- They are generally spherical or pear-shaped with fuzzy edges
- Their diameters range from 1–100 centimetres (0.39–39.37 in), most commonly 10–20 cm (3.9–7.9 in).
- Their brightness corresponds to roughly that of a domestic lamp, so they can be seen clearly in daylight
- A wide range of colours has been observed, red, orange, and yellow being the most common.
- The lifetime of each event is from 1 second to over a minute with the brightness remaining fairly constant during that time
- They tend to move, most often in a horizontal direction at a few metres per second, but may also move vertically, remain stationary or wander erratically.
- Many are described as having rotational motion
- It is rare that observers report the sensation of heat, although in some cases the disappearance of the ball is accompanied by the liberation of heat
- Some display an affinity for metal objects and may move along conductors such as wires or metal fences
- Some appear within buildings passing through closed doors and windows
- Some have appeared within metal aircraft and have entered and left without causing damage
- The disappearance of a ball is generally rapid and may be either silent or explosive
- Odors resembling ozone, burning sulfur, or nitrogen oxides are often reported

Direct Measurements of Natural Ball Lightning

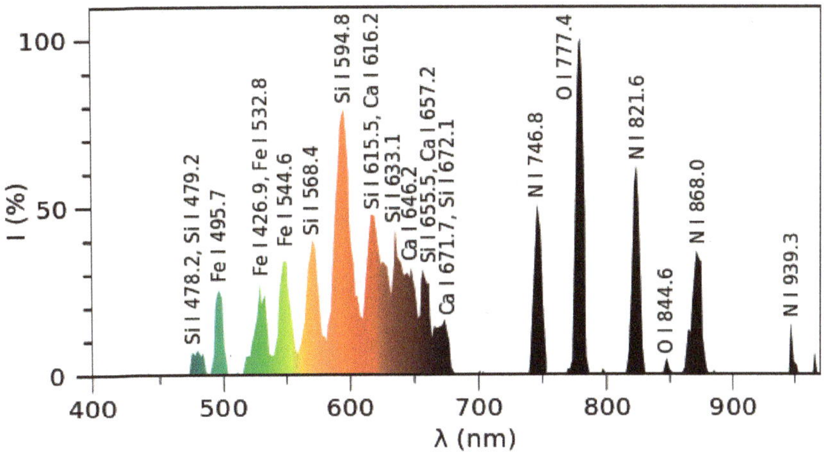

The emission spectrum (intensity vs. wavelength) of a natural ball lightning

In January 2014, scientists from Northwest Normal University in Lanzhou, China, published the results of recordings made in July 2012 of the optical spectrum of what was thought to be natural ball lightning made by chance during the study of ordinary cloud–ground lightning on the Tibetan Plateau. At a distance of 900 m (3,000 ft), a total of 1.64 seconds of digital video of the ball lightning and its spectrum was made, from the formation of the ball lightning after the ordinary lightning struck the ground, up to the optical decay of the phenomenon. Additional video was recorded by a high-speed (3000 frames/sec) camera, which captured only the last 0.78 seconds of the event, due to its limited recording capacity. Both cameras were equipped with slitless spectrographs. The researchers detected emission lines of neutral atomic silicon, calcium, iron, nitrogen and oxygen—in contrast with mainly ionized nitrogen emission lines in the spectrum of the parent lightning. The ball lightning traveled horizontally across the video frame at an average speed equivalent of 8.6 m/s (28 ft/s). It had a diameter of 5 m (16 ft) and covered a distance of about 15 m (49 ft) within those 1.64 s.

Oscillations in the light intensity and in the oxygen and nitrogen emission at a frequency of 100 hertz, possibly caused by the electromagnetic field of the 50 Hz high-voltage power transmission line in the vicinity, were observed. From the spectrum, the temperature of the ball lightning was assessed as being lower than the temperature of the parent lightning (<15,000–30,000 K (14,700–29,700 °C; 26,500–53,500 °F)). The observed data are consistent with vaporization of soil as well as with ball lightning's sensitivity to electric fields.

Laboratory Experiments

Scientists have long attempted to produce ball lightning in laboratory experiments. While some experiments have produced effects that are visually similar to reports of natural ball lightning, it has not yet been determined whether there is any relation.

Nikola Tesla reportedly could artificially produce 1.5-inch (38 mm) balls and conducted some demonstrations of his ability, but he was truly interested in higher voltages and powers, and remote transmission of power, so the balls he made were just a curiosity.

The International Committee on Ball Lightning (ICBL) held regular symposia on the subject. A related group uses the generic name "Unconventional Plasmas". The last ICBL symposium was tentatively scheduled for July 2012 in San Marcos, Texas but was cancelled due to a lack of submitted abstracts.

Wave-Guided Microwaves

Ohtsuki and Ofuruton described producing "plasma fireballs" by microwave interference within an air-filled cylindrical cavity fed by a rectangular waveguide using a 2.45 GHz, 5 kW (maximum power) microwave oscillator.

A demonstration of the water discharge experiment

Water Discharge Experiments

Some scientific groups, including the Max Planck Institute, have reportedly produced a ball lightning-type effect by discharging a high-voltage capacitor in a tank of water.

Home Microwave Oven Experiments

Many modern experiments involve using a microwave oven to produce small rising glowing balls, often referred to as *plasma balls*. Generally, the experiments are conducted by placing a lit or recently extinguished match or other small object in a microwave oven. The burnt portion of the object flares up into a large ball of fire, while "plasma balls" float near the oven chamber ceiling. Some experiments describe covering the match with an inverted glass jar, which contains both the flame and the balls so that they don't damage the chamber walls. (a glass jar, however, eventually explodes rather than simply cause charred paint or melting metal, as happens to the inside of a microwave.) Experiments by Eli Jerby and Vladimir Dikhtyar in Israel revealed that microwave plasma balls are made up of nanoparticles with an average radius of 25 nanometres (9.8×10^{-7} in). The Israeli team demonstrated the phenomenon with copper, salts, water and carbon.

Silicon Experiments

Experiments in 2007 involved shocking silicon wafers with electricity, which vaporizes the silicon and induces oxidation in the vapors. The visual effect can be described as small glowing, sparkling orbs that roll around a surface. Two Brazilian scientists, Antonio Pavão and Gerson Paiva of the Federal University of Pernambuco have reportedly consistently made small long-lasting balls using this method. These experiments stemmed from the theory that ball lightning is actually oxidized silicon vapors *(see vaporized silicon hypothesis, below)*.

Proposed Scientific Explanations

There is at present no widely accepted explanation for ball lightning. Several hypotheses have been advanced since the phenomenon was brought into the scientific realm by the English physician and electrical researcher William Snow Harris in 1843, and French Academy scientist François Arago in 1855.

Vaporized Silicon Hypothesis

This hypothesis suggests that ball lightning consists of vaporized silicon burning through oxidation. Lightning striking Earth's soil could vaporize the silica contained within it, and somehow separate the oxygen from the silicon dioxide, turning it into pure silicon vapor. As it cools, the silicon could condense into a floating aerosol, bound by its charge, glowing due to the heat of silicon recombining with oxygen. An experimental investigation of this effect, published in 2007, reported producing "luminous balls with lifetime in the order of seconds" by evaporating pure silicon with an electric arc. Videos and spectrographs of this experiment have been made available. This hypothesis got significant supportive data in 2014, when the first ever recorded spectra of natural ball lightning were published. The theorized forms of silicon storage in soil include nanoparticles of Si, SiO, and SiC.

Electrically Charged Solid-Core Model

In this model ball lightning is assumed to have a solid, positively charged core. According to this underlying assumption, the core is surrounded by a thin electron layer with a charge nearly equal in magnitude to that of the core. A vacuum exists between the core and the electron layer containing an intense electromagnetic (EM) field, which is reflected and guided by the electron layer. The microwave EM field applies a ponderomotive force (radiation pressure) to the electrons preventing them from falling into the core.

Microwave Cavity Hypothesis

Pyotr Kapitsa proposed that ball lightning is a glow discharge driven by microwave radiation that is guided to the ball along lines of ionized air from lightning clouds where it is produced. The ball serves as a resonant microwave cavity, automatically adjusting its radius to the wavelength of the microwave radiation so that resonance is maintained.

The Handel Maser-Soliton theory of ball lightning hypothesizes that the energy source generating the ball lightning is a large (several cubic kilometers) atmospheric maser. The ball lightning appears as a plasma caviton at the antinodal plane of the microwave radiation from the maser.

Soliton Hypothesis

Julio Rubenstein, David Finkelstein, and James R. Powell proposed that ball lightning is a detached St. Elmo's fire (1964–1970). St. Elmo's fire arises when a sharp conductor, such as a ship's mast, amplifies the atmospheric electric field to breakdown. For a globe the amplification factor is 3. A free ball of ionized air can amplify the ambient field this much by its own conductivity. When this maintains the ionization, the ball is then a soliton in the flow of atmospheric electricity.

Powell's kinetic theory calculation found that the ball size is set by the second Townsend coefficient (the mean free path of conduction electrons) near breakdown. Wandering glow discharges are found to occur within certain industrial microwave ovens and continue to glow for several seconds after power is shut off. Arcs drawn from high-power low-voltage microwave generators also are found to exhibit afterglow. Powell measured their spectra, and found that the after-glow comes mostly from metastable NO ions, which are long-lived at low temperatures. It occurred in air and in nitrous oxide, which possess such metastable ions, and not in atmospheres of argon, carbon dioxide, or helium, which do not.

The soliton model of a ball lightning was further developed. It was suggested that a ball lightning is based on spherically symmetric nonlinear oscillations of charged particles in plasma – the analogue of a spatial Langmuir soliton. These oscillations were described in both classical and quantum approaches. It was found that the most intense plasma oscillations occur in the central regions of a ball lightning. It is suggested that bound states of radially oscillating charged particles with oppositely oriented spins – the analogue of Cooper pairs – can appear inside a ball lightning. This phenomenon, in its turn, can lead to a superconducting phase in a ball lightning. The idea of the superconductivity in a ball lightning was considered earlier. The possibility of the existence of a ball lightning with a composite core was also discussed in this model.

Hydrodynamic Vortex Ring Antisymmetry

Physicist Domokos Tar suggested the following theory for ball lightning formation based on his ball lightning observation. Lightning strikes perpendicular to the ground, and thunder follows immediately at supersonic speed in the form of shock waves that form an invisible aerodynamic turbulence ring horizontal to the ground. Around the ring, over and under pressure systems rotate the vortex around a circular axis in the cross section of the torus. At the same time, the ring expands concentrically parallel to the ground at low speed.

In an open space, the vortex fades and finally disappears. If the vortex's expansion is obstructed, and symmetry is broken, the vortex breaks into cyclical form. Still invisible, and due to the central and surface tension-forces, it shrinks to an intermediate state of a cylinder, and finally into a ball. The resulting transformation subsequently becomes visible once the energy is concentrated into the final spherical stage.

The ball lightning has the same rotational axis as the rotating cylinder. As the vortex has a much smaller vector of energy compared to the overall energy of the reactant sonic shock wave, its vector is likely fractional to the overall reaction. The vortex, during contraction, gives the majority of its energy to form the ball lightning, achieving nominal energy loss.

In some observations, the ball lightning appeared to have an extremely high energy concentration but this phenomenon hasn't been adequately verified. The present theory concerns only the low energy lightning ball form, with centripetal forces and surface tension. The visibility of the ball lightning can be associated with electroluminescence, a direct result of the triboelectric effect from materials within the area of the reaction. Static discharge from the cylindrical stage imply the existence of contact electrification within the object. The direction of the discharges indicate the cylinder's rotation, and resulting rotational axis of the ball lightning in accordance to the law of laminar flow. If the ball came from the channel, it would have rotated in the opposite direction.

One theory that may account for the wide spectrum of observational evidence is the idea of combustion inside the low-velocity region of spherical vortex breakdown of a natural vortex (e.g., the 'Hill's spherical vortex').

Nanobattery Hypothesis

Oleg Meshcheryakov suggests that ball lightning is made of composite nano or submicrometre particles—each particle constituting a battery. A surface discharge shorts these batteries, causing a current that forms the ball. His model is described as an aerosol model that explains all the observable properties and processes of ball lightning.

Black Hole Hypothesis

Another hypothesis is that some ball lightning is the passage of microscopic primordial black holes through the Earth's atmosphere. This possibility was mentioned parenthetically by Leo Vuyk in 1992 in a patent application and a second patent application in 1996 by Leendert Vuyk. The first detailed scientific analysis was published by Mario Rabinowitz in Astrophysics and Space Science journal in 1999. It was inspired by M. Fitzgerald's account of ball lightning on 6 August 1868, in Ireland that left a 6-square-metre (65 sq ft) hole, a 90-metre-long (300 ft) trench, a second trench 25 metres (82 ft) long, and a small cave in the peat bog. Pace VanDevender, a plasma physicist at Sandia National Laboratories in Albuquerque, New Mexico, and his team found depressions consistent with Fitzgerald's report and inferred that the evidence is inconsistent with thermal (chemical or nuclear) and electrostatic effects. An electromagnetically levitated, compact mass of over 20,000 kg (44,000 lb) would produce the reported effects but requires a density of more than 2000 times the density of gold, which implies a miniature black hole. He and his team found a second event in the peat-bog witness plate from 1982 and are trying to geolocate electromagnetic emission consistent with the hypothesis. His colleagues at the institute agreed that, implausible though the hypothesis seemed, it was worthy of their attention.

Buoyant Plasma Hypothesis

The declassified Project Condign report concludes that buoyant charged plasma formations similar to ball lightning are formed by novel physical, electrical, and magnetic phenomena, and that these charged plasmas are capable of being transported at enormous speeds under the influence and balance of electrical charges in the atmosphere. These plasmas appear to originate due to more than one set of weather and electrically charged conditions, the scientific rationale for which is incomplete or not fully understood. One suggestion is that meteors breaking up in the atmosphere and forming charged plasmas as opposed to burning completely or impacting as meteorites could explain some instances of the phenomena, in addition to other unknown atmospheric events.

Transcranial Magnetic Stimulation

Cooray and Cooray (2008) stated that the features of hallucinations experienced by patients having epileptic seizures in the occipital lobe are similar to the observed features of ball lightning. The study also showed that the rapidly changing magnetic field of a close lightning flash is strong enough to excite the neurons in the brain. This strengthens the possibility of lightning-induced seizure in the occipital lobe of a person close to a lightning strike, establishing the connection between epileptic hallucination mimicking ball lightning and thunderstorms.

More recent research with transcranial magnetic stimulation has been shown to give the same hallucination results in the laboratory (termed magnetophosphenes), and these conditions have been shown to occur in nature near lightning strikes. This hypothesis fails to explain observed physical damage caused by ball lightning or simultaneous observation by multiple witnesses. (At the very least, observations would differ substantially.)

Theoretical calculations from University of Innsbruck researchers suggest that the magnetic fields involved in certain types of lightning strikes could potentially induce visual hallucinations resembling ball lightning. Such fields, which are found within close distances to a point in which multiple lightning strikes have occurred over a few seconds, can directly cause the neurons in the visual cortex to fire, resulting in magnetophosphenes (magnetically induced visual hallucinations).

Spinning Plasma Toroid (Ring)

Seward proposes that ball lightning is a spinning plasma toroid or ring. He built a lab that produces lightning level arcs, and by modifying the conditions he produced bright, small balls that mimic ball lightning and persist in atmosphere after the arc ends. Using a high speed camera he was able to show that the bright balls were spinning plasma toroids.

Chen was able to derive the physics and found that there is a class of plasma toroids that remain stable with or without an external magnetic containment, a new plasma configuration unlike anything reported elsewhere.

Seward published images of the results of his experiments, along with his method. Included is a report by a farmer of observing a ball lightning event forming in a kitchen and the effects it caused as it moved around the kitchen. This is the only eye witness account of ball lightning forming, then staying in one area, then ending that the author has heard of.

Rydberg Matter Concept

Manykin et al. have suggested atmospheric Rydberg matter as an explanation of ball lightning phenomena. Rydberg matter is a condensed form of highly excited atoms in many aspects similar to electron-hole droplets in semiconductors. However, in contrast to electron-hole droplets, Rydberg matter has an extended life-time—as long as hours. This condensed excited state of matter is supported by experiments, mainly of a group led by Holmlid. It is similar to a liquid or solid state of matter with extremely low (gas-like) density. Lumps of atmospheric Rydberg matter can result from condensation of highly excited atoms that form by atmospheric electrical phenomena, mainly due to linear lightning. Stimulated decay of Rydberg matter clouds can, however, take the form of an avalanche, and so appear as an explosion.

Other Hypotheses

Several other hypotheses have been proposed to explain ball lightning:

- Spinning electric dipole hypothesis. A 1976 article by V. G. Endean postulated that ball lightning could be described as an electric field vector spinning in the microwave frequency region.

- Electrostatic Leyden jar models. Stanley Singer discussed (1971) this type of hypothesis and suggested that the electrical recombination time would be too short for the ball lightning lifetimes often reported.
- J. Pace VanDevender separates extreme ball lightning of the highly energetic violent kind, and proposes a theory of neutrinos and heavy neutrinos.
- Smirnov proposed (1987) a fractal aerogel hypothesis.
- V.P. Torchigin proposed (2003) considering ball lightning as a form of self-confined intense light.
- M.I. Zelikin proposed (2006) an explanation (with a rigorous mathematical foundation) based on the hypothesis of plasma superconductivity .
- H.C. Wu proposed (2016) that ball lightning arises when a relativistic electron bunch forming at the tip of a lightning stroke excites intense microwave radiation under certain conditions. As the microwaves ionize the surrounding air, their associated pressure may then evacuate the resulting plasma to form a bubble that stably traps the radiation.

REFERENCES

1. Likens, Gene E.; Keene, William C.; Miller, John M.; Galloway, James N. (1987). “Chemistry of precipitation from a remote, terrestrial site in Australia”. Journal of Geophysical Research. 92: 13299. doi:10.1029/JD092iD11p13299.
2. E. S. de Beer, ed. The Diary of John Evelyn, III, 1955 (19 September 1667) p. 495.
3. Glossary, United States: NASA Earth Observatory, acid rain, archived from the original on 2011-12-13, retrieved 2013-02-15
4. Weathers, K. C. and Likens, G. E. (2006). “Acid rain”, pp. 1549–1561 in: W. N. Rom and S. Markowitz (eds.). Environmental and Occupational Medicine. Lippincott-Raven Publ., Philadelphia. Fourth Edition, ISBN 0-7817-6299-5.
5. Seinfeld, John H.; Pandis, Spyros N (1998). Atmospheric Chemistry and Physics — From Air Pollution to Climate Change. John Wiley and Sons, Inc. ISBN 978-0-471-17816-3
6. Likens, G. E.; Bormann, F. H.; Johnson, N. M. (1972). “Acid rain”. Environment. 14 (2): 33–40. doi:10.1080/00139157.1972.9933001.
7. Acid Rain in New England, A Brief History ArchivedSeptember 25, 2010, at the Wayback Machine.. Epa.gov. Retrieved on 2013-02-09.
8. Likens, G. E.; Bormann, F. H. (1974). “Acid Rain: A Serious Regional Environmental Problem”. Science. 184 (4142): 1176–9. doi:10.1126/science.184.4142.1176. PMID 17756304.
9. Keller, C. K.; White, T. M.; O’Brien, R.; Smith, J. L. (2006). “Soil CO2 dynamics and fluxes as affected by tree harvest in an experimental sand ecosystem”. Journal of Geophysical Research. 111. doi:10.1029/2005JG000157.

10. Galloway, JN; Dianwu, Z; Jiling, X; Likens, GE (1987). "Acid rain: China, United States, and a remote area". Science. 236 (4808): 1559–62. doi:10.1126/science.236.4808.1559. PMID 17835740.
11. Chandru (2006-09-09). "CHINA: Industrialization pollutes its country side with Acid Rain". Southasiaanalysis.org. Archived from the original on June 20, 2010. Retrieved 2010-11-18.
12. Lefohn, A.S.; Husar, J.D.; Husar, R.B. (1999), Global Sulfur Emissions Database, United States: A.S.L. & Associates
13. Likens, G. E.; Wright, R. F.; Galloway, J. N.; Butler, T. J. (1979). "Acid rain". Scientific American. 241 (4): 43–51. doi:10.1038/scientificamerican1079-43.
14. Likens, G. E. (1984). "Acid rain: the smokestack is the "smoking gun"". Garden. 8 (4): 12–18.
15. "Art Under Wraps", Harvard Magazine, March–April 2000
16. Likens, Gene E.; Bormann, F. Herbert; Johnson, Noye M. "Acid Rain". Environment: Science and Policy for Sustainable Development. 14 (2): 33–40. doi:10.1080/00139157.1972.9933001.
17. Johnson, Noye M.; Driscoll, Charles T.; Eaton, John S.; Likens, Gene E.; McDowell, William H. (1981-09-01). "'Acid rain', dissolved aluminum and chemical weathering at the Hubbard Brook Experimental Forest, New Hampshire". Geochimica et Cosmochimica Acta. 45 (9): 1421–1437. doi:10.1016/0016-7037(81)90276-3.
18. Hall, Ronald J.; Likens, Gene E.; Fiance, Sandy B.; Hendrey, George R. (1980-08-01). "Experimental Acidification of a Stream in the Hubbard Brook Experimental Forest, New Hampshire". Ecology. 61 (4): 976–989. doi:10.2307/1936765. ISSN 1939-9170.
19. Lackey, R.T. (1997). "Science, policy, and acid rain: lessons learned" (PDF). Renewable Resources Journal. 15 (1): 9–13.
20. Winstanley, Derek; Lackey, Robert T.; Warnick, Walter L.; Malanchuk, John (1998). "Acid rain: Science and policy making". Environmental Science & Policy. 1: 51. doi:10.1016/S1462-9011(98)00006-9.
21. Times, Robert Reinhold, Special To The New York (1982-06-08). "ACID RAIN ISSUE CREATES STRESS BETWEEN ADMINISTRATION AND SCIENCE ACADEMY". The New York Times. ISSN 0362-4331. Retrieved 2016-11-16.
22. "Ronald Reagan on Environment". www.ontheissues.org. Retrieved 2016-11-16.
23. "HYSTERIA ABOUT ACID RAIN Even Ronald Reagan now casts it as the villain. He is overriding a lot of scientific evidence. - April 14, 1986". archive.fortune.com. Retrieved 2016-11-16.
24. "Ronald Reagan: Nomination of William A. Nierenberg To Be a Member of the National Science Board". www.presidency.ucsb.edu. Retrieved 2016-11-16.
25. "Report of the Acid Rain Peer Review Panel". Document Display | NEPIS | US EPA. July 1984. Retrieved 2016-11-16.

26. "From tobacco to climate change, 'merchants of doubt' undermined the science". Grist. 2010-04-17. Retrieved 2016-11-16.
27. Franklin, Ben A. (1984-08-18). "LEGISLATORS SAT WHITE HOUSE SUPPRESSED ACID RAIN REPORT". The New York Times. ISSN 0362-4331. Retrieved 2016-11-16.
28. "US EPA: A Brief History of Acid Rain". Epa.gov. Archived from the original on 2010-09-25. Retrieved 2010-11-18.
29. 'Cap-and-trade' model eyed for cutting greenhouse gases, San Francisco Chronicle, December 3, 2007.
30. "Facts On File News Services Databases". 2facts.com. Retrieved 2010-11-18.
31. Gilberston, T. and Reyes, O. 2009. Carbon Trading: how it works and why it fails. Dag Hammerskjold Foundation: 22
32. Acid Rain Program 2007 Progress Report, United States Environmental Protection Agency, January 2009.
33. Gerdes, Justin. "Cap and Trade Curbed Acid Rain: 7 Reasons Why It Can Do The Same For Climate Change". Forbes. Forbes. Retrieved 27 Oct 2014.
34. Muki Haklay (2015). "Citizen Science and Policy: A European Perspective" (PDF). Woodrow Wilson International Center for Scholars. p. 11. Retrieved 3 June 2016.
35. R. Kerson (1989). "Lab for the Environnment". MIT Technology Review. Vol. 92 no. 1. pp. 11–12.
36. Berresheim, H.; Wine, P.H. and Davies D.D. (1995). "Sulfur in the Atmosphere". In Composition, Chemistry and Climate of the Atmosphere, ed. H.B. Singh. Van Nostrand Rheingold ISBN 0-442-01264-0
37. Clean Air Act Reduces Acid Rain In Eastern United States, ScienceDaily, September 28, 1998
38. "UK National Air Quality Archive: Air Pollution Glossary". Airquality.co.uk. 2002-04-01. Retrieved 2010-11-18.
39. US EPA: Effects of Acid Rain – Surface Waters and Aquatic Animals
40. Rodhe, H., et al. The global distribution of acidifying wet deposition. Environmental Science and TEchnology. vlo. 36, no. 20 (October) p. 4382–8
41. US EPA: Effects of Acid Rain – Forests Archived July 26, 2008, at the Wayback Machine.
42. Likens, G.E.; Driscoll, C.T.; Buso, D.C.; Mitchell, M.J.; Lovett, G.M.; Bailey, S.W.; Siccama, T.G.; Reiners, W.A.; Alewell, C. (2002). "The biogeochemistry of sulfur at Hubbard Brook" (PDF). Biogeochemistry. 60 (3): 235. doi:10.1023/A:1020972100496.
43. Likens, G. E.; Driscoll, C. T.; Buso, D. C. (1996). "Long-Term Effects of Acid Rain: Response and Recovery of a Forest Ecosystem" (PDF). Science. 272 (5259): 244. doi:10.1126/science.272.5259.244.

44. DeHayes, D.H., Schaberg, P.G. and G.R. Strimbeck. (2001). Red Spruce Hardiness and Freezing Injury Susceptibility. In: F. Bigras, ed. Conifer Cold Hardiness. Kluwer Academic Publishers, the Netherlands ISBN 0-7923-6636-0.
45. Lazarus, Brynne E.; Schaberg, Paul G.; Hawley, Gary J.; DeHayes, Donald H. (2006). "Landscape-scale spatial patterns of winter injury to red spruce foliage in a year of heavy region-wide injury" (PDF). Can. J. For. Res. 36: 142–152. doi:10.1139/x05-236. highbeam copy
46. Effects of Acid Rain – Human Health. Epa.gov (2006-06-02). Retrieved on 2013-02-09.
47. "ICP on effects on materials". Springerlink.com. Retrieved 2010-11-18.
48. "Approaches in modeling the impact of air pollution-induced material degradation" (PDF). Retrieved 2010-11-18.
49. Ed. Hatier (1993). "Acid Rain in Europe". United Nations Environment Programme GRID Arendal. Retrieved 2010-01-31.
50. US Environmental Protection Agency (2008). "Clean Air Markets 2008 Highlights". Retrieved 2010-01-31.
51. Clean Air Act Amendments of 1990, 42 U.S. Code 7651
52. Halldór Björnsson. "Global circulation". Retrieved September 2, 2016. Veðurstofa Íslands. Retrieved on 2008-06-15.
53. Polar vortex. Glossary of Meteorology. American Meteorological Society. June 2000. Retrieved 15 June 2008.
54. Stratospheric Polar Vortex Influences Winter Cold, Researchers Say (Press release). American Association for the Advancement of Science. December 3, 2001. Retrieved May 23, 2015.
55. Air Maps, Littell›s Living Age No. 495, 12 November 1853, p. 430.
56. GEOS-5 Analyses and Forecasts of the Major Stratospheric Sudden Warming of January 2013 (Press release). Goddard Space Flight Center. Retrieved January 8, 2014.
57. http://blog.quarkexpeditions.com/polar-vortex-the-science-myth-media-hype-behind-north-american-weather-phenomen.
58. Hartmann, D; Schoeberl, M (1991). "Mixing of polar vortex air into middle latitudes as revealed by tracer-tracer scatterplots". Journal of Geophysical Research. 102: 13119. Bibcode:1997JGR...10213119W. doi:10.1029/96JD03715.
59. Cavallo, Steven M.; Hakim, Gregory J. (April 2009). "Potential Vorticity Diagnosis of a Tropopause Polar Cyclone". Monthly Weather Review. 137 (4): 1358–1371. Bibcode:2009MWRv..137.1358C. doi:10.1175/2008MWR2670.1.
60. Kolstad, Erik W.; Breiteig, Tarjei; Scaife, Adam A. (April 2010). "The association between stratospheric weak polar vortex events and cold air outbreaks in the Northern Hemisphere". Quarterly Journal of the Royal Meteorological Society. Royal Meteorological Society. 136: 887. Bibcode:2010EGUGA..12.5739K. doi:10.1002/qj.620.

61. Abdolreza Kashki & Javad Khoshhal (2013-11-22). "Investigation of the Role of Polar Vortex in Iranian First and Last Snowfalls". Journal of Geology and Geography. Canadian Center of Science and Education. 5 (4). ISSN 1916-9779.

62. Erik A. Rasmussen and John Turner (2003). Polar lows: mesoscale weather systems in the polar regions. Cambridge University Press. p. 174. ISBN 978-0-521-62430-5.

63. Robock, Alan (2000). "Volcanic eruptions and climate". Reviews of Geophysics. 38 (2): 191–219. Bibcode:2000RvGeo..38..191R. doi:10.1029/1998RG000054.

64. Todd Mitchell (2004). Arctic Oscillation (AO) time series, 1899 – June 2002. University of Washington. Retrieved on 2009-03-02.

65. Kevin Myatt (2005-01-17). Cold enough for snow, and more's on the way. Roanoke Times. Retrieved on 2012-02-24.

66. Nash, E; Newman, P; Rosenfield, J; Schoeberl, M (2012). "An objective determination of the polar vortex using Ertel's potential vorticity". Journal of Geophysical Research. 101 (D5): 9471–9478. Bibcode:1996JGR...101.9471N. doi:10.1029/96JD00066.

67. Li, L; Li, C; Pan, Y (2012). "On the differences and climate impacts of early and late stratospheric polar vortex breakup". Advances in Atmospheric Sciences. 29 (5): 1119–1128. Bibcode:2012AdAtS..29.1119L. doi:10.1007/s00376-012-1012-4.

68. Wei, K; Chen, W; Huang, R (2007). "Dynamical diagnosis of the breakup of the stratospheric polar vortex in the northern hemisphere". Science in China Series D (Earth Sciences). 50 (9): 1369–1379. doi:10.1007/s11430-007-0100-2.

69. Reichler, Tom; Kim, J; Manzini, E; Kroger, J (2012). "A stratospheric connection to Atlantic climate variability". Nature Geoscience. 5: 783–787. Bibcode:2012NatGe...5..783R. doi:10.1038/ngeo1586.

70. Ripesi, Patrizio; et al. (2012). "The February 2010 Artcic Oscillation Index and its stratospheric connection" (PDF). Quarterly Journal of the Royal Meteorological Society. 138: 1961–1969. Bibcode:2012QJRMS.138.1961R. doi:10.1002/qj.1935.

71. Reichler, Tom; Kim, J; Manzini, E; Kroger, J (2012). "A stratospheric connection to Atlantic climate variability". Nature Geoscience. 5: 783–787. Bibcode:2012NatGe...5..783R. doi:10.1038/ngeo1586.

72. Limpasuvan, Varavut; Hartmann, Dennis L.; Thompson, David W. J.; Jeev, Kumar; Yung, Yuk L. (2005). "Stratosphere-troposphere evolution during polar vortex intensification"(PDF). Journal of Geophysical Research. 110 (D24): 27. Bibcode:2005JGRD..11024101L. doi:10.1029/2005JD006302.

73. Cavallo, S; Hakim, G.J. (2013). "Physical mechanisms of tropopause polar vortex intensity change". Journal of the Atmospheric Sciences. 70 (11): 3359–3373. doi:10.1175/JAS-D-13-088.1.

74. Hartmann, D; Schoeberl, M (1991). "The dynamics of the stratospheric polar vortex and its relation to springtime ozone depletions". Science. 251 (4989): 46–52. Bibcode:1991Sci...251...46S. doi:10.1126/science.251.4989.46. PMID 17778602.

75. Widnall, S; Sullivan, J (1973). "On the stability of vortex rings". Proceedings of the Royal Society of London. Series A, Mathematical and Physical Sciences. 332 (1590): 335–353. Bibcode:1973RSPSA.332..335W. doi:10.1098/rspa.1973.0029.

76. Manney, G; Zurek, R; O'Neill, A; Swinbank, R (1994). "On the motion of air through the stratospheric polar vortex". Journal of the Atmospheric Sciences. 51 (20): 2973–2994. Bibcode:1994JAtS...51.2973M. doi:10.1175/1520-0469(1994)051<2973:otmoat>2.0.co;2.

77. Waugh, D; Plumb, R; Elkins, J; Fahey, D; Boering, K; Dutton, G; Lait, L (2012). "Mixing of polar vortex air into middle latitudes as revealed by tracer-tracer scatterplots". Journal of Geophysical Research: Atmospheres. 102 (D11): 13199–1314. Bibcode:1997JGR...10213119W. doi:10.1029/96JD03715.

78. "Archived copy". Archived from the original on 2015-12-07. Retrieved 2015-11-26.

79. Erdman, Jon (2014). "What's a Polar Vortex?: The Science Behind Arctic Outbreaks". wunderground. Retrieved 25 February 2014.

80. Baldwin, M. P.; Dunkerton, TJ (2001). "Stratospheric Harbingers of Anomalous Weather Regimes". Science. 294 (5542): 581–4. Bibcode:2001Sci...294..581B. doi:10.1126/science.1063315. PMID 11641495.

81. NASA (December 21, 2001). "Stratospheric Polar Vortex Influences Winter Cold". Earth Observatory. Retrieved January 7, 2014.

82. Song, Yucheng; Robinson, Walter A. (2004). "Dynamical Mechanisms for Stratospheric Influences on the Troposphere". Journal of the Atmospheric Sciences. 61 (14): 1711–25. Bibcode:2004JAtS...61.1711S. doi:10.1175/1520-0469(2004)061<1711:DMFSIO>2.0.CO;2.

83. Overland, James E. (2013). "Atmospheric science: Long-range linkage". Nature Climate Change. 4: 11–2. Bibcode:2014NatCC...4...11O. doi:10.1038/nclimate2079.

84. Tang, Qiuhong; Zhang, Xuejun; Francis, Jennifer A. (2013). "Extreme summer weather in northern mid-latitudes linked to a vanishing cryosphere". Nature Climate Change. 4: 45–50. Bibcode:2014NatCC...4...45T. doi:10.1038/nclimate2065.

85. Screen, J A (2013). "Influence of Arctic sea ice on European summer precipitation". Environmental Research Letters. 8 (4): 044015. Bibcode:2013ERL.....8d4015S. doi:10.1088/1748-9326/8/4/044015.

86. Francis, Jennifer A.; Vavrus, Stephen J. (2012). "Evidence linking Arctic amplification to extreme weather in mid-latitudes". Geophysical Research Letters. 39 (6): n/a. Bibcode:2012GeoRL..39.6801F. doi:10.1029/2012GL051000.

87. Petoukhov, Vladimir; Semenov, Vladimir A. (2010). "A link between reduced Barents-Kara sea ice and cold winter extremes over northern continents". Journal of Geophysical Research. 115. Bibcode:2010JGRD..11521111P. doi:10.1029/2009JD013568.

88. Masato, Giacomo; Hoskins, Brian J.; Woollings, Tim (2013). "Winter and Summer Northern Hemisphere Blocking in CMIP5 Models". Journal of Climate. 26 (18): 7044–59. doi:10.1175/JCLI-D-12-00466.1.
89. Friedlander, Blaine (March 4, 2013). "Arctic ice loss amplified Superstorm Sandy violence". Cornell Chronicle.
90. J. A. Pyle (1997-04-08). The Arctic and environmental change. CRC Press. pp. 42–44. ISBN 978-90-5699-020-6.
91. Rolf Müller (2010). Tracer-tracer Relations as a Tool for Research on Polar Ozone Loss. Forschungszentrum Jülich. p. 47. ISBN 978-3-89336-614-9.
92. K. Mohanakuma (2008). Stratosphere troposphere interactions: an introduction. Springer. p. 34. ISBN 978-1-4020-8216-0.
93. "Double vortex at Venus South Pole unveiled". European Space Agency. Retrieved June 2006. Check date values in: |access-date= (help)
94. "Saturn's Bull's-Eye Marks Its Hot Spot". NASA. 2005. Retrieved January 8, 2014.
95. J. B, ed. (1905). The Two Widecombe Tracts, 1638 giving a Contemporary Account of the great Storm, reprinted with an Introduction. Exeter: James G Commin. Retrieved 29 June 2013.
96. Day, Jeremiah (January 1813). "A view of the theories which have been proposed to explain the origin of meteoric stones". The General Repository and Review. Cambridge, Massachusetts: William Hilliard. 3 (1): 156–157. Retrieved 29 June 2013.
97. How Stuff Works entry. Retrieved 21 January 2014.
98. Anna Salleh (2008-03-20). "Ball lightning bamboozles physicist". 35.2772;149.1292: Abc.net.au. Retrieved 21 January 2014.
99. Cen, Jianyong; Yuan, Ping; Xue, Simin (17 January 2014). "Observation of the Optical and Spectral Characteristics of Ball Lightning". Physical Review Letters. American Physical Society. 112 (3): 035001. Bibcode:2014PhRvL.112c5001C. doi:10.1103/PhysRevLett.112.035001. Retrieved 19 January 2014.
100. Slezak, Michael (16 January 2014). "Natural ball lightning probed for the first time". New Scientist. Retrieved 22 January 2014.
101. Anon. "Ask the experts". Scientific American. Retrieved 4 April 2007.
102. McNally, J. R. (1960). "Preliminary Report on Ball Lightning". Proceedings of the Second Annual Meeting of the Division of Plasma Physics of the American Physical Society (Paper J-15 ed.). Gatlinburg. pp. 1–25.
103. Grigoriev, A. I. (1988). Y. H. Ohtsuki, ed. "Statistical Analysis of the Ball Lightning Properties". Science of Ball Lightning. Singapore: World Scientific Publishing Co.: 88–134.

104. Anon. "Foreign Affairs: Bristol 17 December". Weekly Journal or British Gazetteer. 24 December 1726.

105. Anon (24 December 1726). "Foreign Affairs: London 24 December". London Journal.

106. Clarke, Ronald W. (1983). Benjamin Franklin, A Biography. Random House. p. 87. ISBN 978-1-84212-272-3. Archived from the original on 12 January 2012.

107. Simons, Paul (17 February 2009). "Weather Eye Charles Darwin the meteorologist". The Times. London. Retrieved 16 April 2010.

108. http://www.thenational.ae/article/20090223/FRONTIERS/646186738/1036

109. Brewer, Ebenezer Cobham (1864). A Guide to the Scientific Knowledge of Things Familiar. pp. 13–14. Retrieved 22 January 2014.

110. de Fonvielle, Wilfrid (1875). "Chapter X Globular lightning". Thunder and lightning (full text). translated by T L Phipson. pp. 32–39. ISBN 978-1-142-61255-9. Retrieved 22 January 2014.

111. Anon (24 December 1867). "Globular lightning". The Leeds mercury. Leeds, UK.

112. "Tsar-Martyr Nicholas II And His Family". Orthodox.net. Archived from the original on 17 June 2009. Retrieved 13 July 2009.

113. There is no present-day Lake Pasquaney in New Hampshire, United States. New Hampshire's Newfound Lake has a Camp Pasquaney. However, part of the lake is known as Pasquaney Bay.

114. Crowley, Aleister (5 December 1989). "Chp. 83". The Confessions of Aleister Crowley: An Autobiography. Penguin. ISBN 0-14-019189-5.

115. Jennison, R. C. (1969). "Ball Lightning". Nature. 224 (5222): 895. Bibcode:1969Natur.224..895J. doi:10.1038/224895a0.

116. "Miracle saved panth". Sikhnet.com. 2009-12-21. Retrieved 21 January 2014.

117. Golden Globe, 24 November 1894.

118. Mark Stenhoff, Ball Lightning: An Unsolved Problem in Atmospheric Physics (Springer Science & Business Media, 2006) p70

119. "The Cape Naturaliste Lighthouse". Lighthouses of Western Australia. Lighthouses of Australia Inc. Retrieved 13 July 2009.

120. Wilder, Laura Ingalls (1937). On the Banks of Plum Creek. Harper Trophy. ISBN 0-06-440005-0.

121. Getline, Meryl (17 October 2005). "Playing with (St. Elmo's) fire". USA Today.

122. "Ball lightning – and the charge sheath vortex". Peter-thomson.co.uk. Retrieved 13 July 2009.

123. This may be an incorrect translation of the word "blixtlokaliseringssystem" from the university article cited in the sources

124. Larsson, Anders (23 April 2002). "Ett fenomen som gäckar vetenskapen" (in Swedish). Uppsala University. Retrieved 19 November 2007.

125. "Fizikai Szemle 2004/10". Kfki.hu. Retrieved 21 January 2014.

126. Domokos Tar (2009). "Observation of Lightning Ball (Ball Lightning): A new phenomenological description of the phenomenon". Proceedings of the 9th International Symposium on Ball Lightning, Aug. Eindhoven. 0910: 783. arXiv:0910.0783 . Bibcode:2009arXiv0910.0783T.

127. Domokos Tar (2010). "Lightning Ball (Ball Lightning) Created by Thunder, Shock-Wave". arXiv:1007.3348 .

128. Domokos Tar (2009). "New Revelation of Lightning Ball Observation and Proposal for a Nuclear Reactor Fusion Experiment". Proceedings 10th International Symposium on Ball Lightning (ISBL-08), July 7–12 Kaliningrad, Russia, pp. 135–141, Eds. Vladimir L. Bychkov & Anatoly I. Nikitin. 0910: 2089. arXiv:0910.2089 . Bibcode:2009arXiv0910.2089T.

129. "Source (in Czech)". Zpravy.idnes.cz. Retrieved 21 January 2014.

130. "The Aviation Herald". avherald.com.

131. "BL_Info_10". Ernmphotography.com. Retrieved 13 July 2009.

132. "Unusual Phenomea Reports: Ball Lightning". Amasci.com. Retrieved 13 July 2009.

133. Barry, James Dale: Ball lightning and bead lightning: extreme forms of atmospheric electricity, ISBN 0-306-40272-6, 1980, Plenum Press (p. 35)

134. Barry, J.D. (1980a) Ball Lightning and Bead Lightning: Extreme Forms of Atmospheric Electricity. 8–9. New York and London: Plenum Press. ISBN 0-306-40272-6

135. Charman, Neil (14 December 1972). "The enigma of ball Lightning". New Scientist. Reed Business information. 56 (824): 632–635. Retrieved 10 May 2010.

136. Ball, Philip (17 January 2014). "Focus: First Spectrum of Ball Lightning". Focus. American Physical Society. doi:10.1103/Physics.7.5. Retrieved 19 January 2014.

137. "The New Wizard of the West". Homepage.ntlworld.com. Retrieved 13 July 2009.

138. Tesla, Nikola (1978). Nikola Tesla – Colorado Springs Notes 1899–1900. Nolit (Beograd, Yugoslavia), 368–370. ISBN 978-0-913022-26-9

139. Anon (2008). "Tenth international syposium on ball lightning/ International symposium III on unconventional plasmas". ICBL. Retrieved 10 May 2010.

140. https://web.archive.org/web/20120604074802/http://www.cose.txstate.edu/isbl_12.html. Archived from the original on 4 June 2012. Retrieved 4 June 2012. Missing or empty |title=(help)

141. Ohtsuki, Y. H.; H. Ofuruton (1991). "Plasma fireballs formed by microwave interference in air". Nature. 350 (6314): 139–141. Bibcode:1991Natur.350..139O. doi:10.1038/350139a0.

142. Ohtsuki, Y. H.; H. Ofuruton (1991). "Plasma fireballs formed by microwave interference in air (Corrections)". Nature. 353 (6347): 868. Bibcode:1991Natur.353..868O. doi:10.1038/353868a0.

143. "'Ball lightning' created in German laboratory | COSMOS magazine". COSMOS magazine. 7 June 2006. Retrieved 13 July 2009.

144. Youichi Sakawa; Kazuyoshi Sugiyama; Tetsuo Tanabe; Richard More (12 July 2006). "Fireball Generation in a Water Discharge". Plasma and Fusion Research: Rapid Communications. Retrieved 13 July 2009.

145. "How to make a Stable Plasmoid (Ball Lightning) with the GMR (Graphite Microwave Resonator) by Jean-Louis Naudin". Jlnlabs.online.fr. 22 December 2005. Archivedfrom the original on 26 June 2009. Retrieved 13 July 2009.

146. "Creating the 4th state of matter with microwaves by Halina Stanley". scienceinschool.org. 13 August 2009. Archivedfrom the original on 31 October 2009. Retrieved 6 October 2009.

147. "Universidade Federal de Pernambuco". Ufpe.br. Archived from the original on 21 June 2009. Retrieved 13 July 2009.

148. "Pesquisadores da UFPE geram, em laboratório, fenômeno atmosférico conhecido como bolas luminosas". Ufpe.br. 16 January 2007. Archived from the original on 20 December 2008. Retrieved 13 July 2009.

149. "Ball Lightning Mystery Solved? Electrical Phenomenon Created in Lab". News. nationalgeographic.com. 21 November 2005. Retrieved 13 July 2009.

150. Snow Harris, William (2008). "Section I". On the nature of thunderstorms (originally published in 1843) (Reprint ed.). Bastian Books. pp. 34–43. ISBN 0-554-87861-5. Retrieved 11 May 2010.

151. François Arago, Meteorological Essays by , Longman, 1855

152. Paiva, Gerson Silva; Antonio Carlos Pavão; Elder Alpes de Vasconcelos; Odim Mendes, Jr.; Eronides Felisberto da Silva, Jr. (2007). "Production of Ball-Lightning-Like Luminous Balls by Electrical Discharges in Silicon". Phys. Rev. Lett. 98 (4): 048501. Bibcode:2007PhRvL..98d8501P. doi:10.1103/PhysRevLett.98.048501. PMID 17358820.

153. "Lightning balls created in the lab". New Scientist. 10 January 2007. A more down-to-earth theory, proposed by John Abrahamson and James Dinniss at the University of Canterbury in Christchurch, New Zealand, is that ball lightning forms when lightning strikes soil, turning any silica in the soil into pure silicon vapour. As the vapour cools, the silicon condenses into a floating aerosol bound into a ball by charges that gather on its surface, and it glows with the heat of silicon recombining with oxygen.

154. "Ball Lightning Mystery Solved? Electrical Phenomenon Created in Lab". National Geographic News. 22 January 2007.

155. ftp://ftp.aip.org/epaps/phys_rev_lett/E-PRLTAO-98-047705/

156. Slezak, Michael. "Natural ball lightning probed for the first time". New Scientist. Retrieved 17 January 2014.

157. Abrahamson, John; Dinniss, James (2000). "Ball lightning caused by oxidation of nanoparticle networks from normal lightning strikes on soil". Nature. 403 (6769): 519–21. Bibcode:2000Natur.403..519A. doi:10.1038/35000525. PMID 10676954.

158. Muldrew, D. B. (1990). "The Physical Nature of Ball Lightning". Geophysical Research Letters. 17 (12): 2277–2280. Bibcode:1990GeoRL..17.2277M. doi:10.1029/GL017i012p02277.

159. Muldrew, D. B. (2010). "Solid charged-core model of ball lightning". Annales Geophysicae. 28: 223–2010. Bibcode:2010AnGeo..28..223M. doi:10.5194/angeo-28-223-2010.

160. Капица, П. Л. (1955). О природе шаровой молнии. Докл. Акад. наук СССР (in Russian). 101: 245.

161. Kapitsa, Peter L. (1955). The Nature of Ball Lightning. In Donald J. Ritchie. Ball Lightning: A Collection of Soviet Research in English Translation (1961 ed.). Consultants Bureau, New York. pp. 11–16. Retrieved 22 January 2014.

162. Handel, Peter H.; Jean-François Leitner (1994). Development of the maser-caviton ball lightning theory. J. Geophys. Res. 99: 10689. Bibcode:1994JGR....9910689H. doi:10.1029/93JD01021.

163. Dvornikov, Maxim; Dvornikov, Sergey (2006). Gerard, F., ed. Electron gas oscillations in plasma. Theory and applications. Advances in Plasma Physics Research. New York, USA: Nova Science Publishers, Inc.: 197–212. arXiv:physics/0306157 . Bibcode:2003physics...6157D. ISBN 1-59033-928-2.

164. Dvornikov, Maxim (2010). Formation of bound states of electrons in spherically symmetric oscillations of plasma. Physica Scripta. 81 (5): 055502. arXiv:1002.0764 . Bibcode:2010PhyS...81e5502D. doi:10.1088/0031-8949/81/05/055502.

165. Dvornikov, Maxim (2011-12-01). Axially and spherically symmetric solitons in warm plasma. Journal of Plasma Physics. 77 (6): 749–764. arXiv:1010.0701 . Bibcode:2011JPlPh..77..749D. doi:10.1017/S002237781100016X. ISSN 1469-7807.

166. Dvornikov, Maxim (2012-02-08). Effective attraction between oscillating electrons in a plasmoid via acoustic wave exchange. Proc. R. Soc. A. 468 (2138): 415–428. arXiv:1102.0944 . doi:10.1098/rspa.2011.0276. ISSN 1364-5021.

167. Dvornikov, Maxim (2012-11-01). Quantum exchange interaction of spherically symmetric plasmoids. Journal of Atmospheric and Solar-Terrestrial Physics. 89: 62–66. arXiv:1112.0239 . Bibcode:2012JASTP..89...62D. doi:10.1016/j.jastp.2012.08.005.

168. Dvornikov, Maxim (2013). Pairing of charged particles in a quantum plasmoid. Journal of Physics A: Mathematical and Theoretical. 46 (4): 045501. arXiv:1208.2208 . Bibcode:2013JPhA...46d5501D. doi:10.1088/1751-8113/46/4/045501.

169. Dvornikov, M. (2013). Stable Langmuir solitons in plasma with diatomic ions. Nonlinear Processes in Geophysics. 20 (4): 581–588. arXiv:1203.0258 . Bibcode:2013NPGeo..20..581D. doi:10.5194/npg-20-581-2013.

170. Davydova, T. A.; Yakimenko, A. I.; Zaliznyak, Yu. A. (2005-02-28). Stable spatial Langmuir solitons. Physics Letters A. 336 (1): 46–52. Bibcode:2005PhLA..336...46D. doi:10.1016/j.physleta.2004.11.063.

171. Dijkhuis, G. C. (1980-03-13). A model for ball lightning. Nature. 284 (5752): 150–151. Bibcode:1980Natur.284..150D. doi:10.1038/284150a0.

172. Zelikin, M. I. (2008). Superconductivity of plasma and fireballs. Journal of Mathematical Sciences. 151 (6): 3473–3496. doi:10.1007/s10958-008-9047-x.

173. Tar, Domokos (2009-10-05). Observation of Lightning Ball (Ball Lightning): A new phenomenological description of the phenomenon. Proceedings of the -th International Symposium on Ball Lightning, Aug. Eindhoven. 9 (2006). arXiv:0910.0783 .

174. Tar, Domokos (2010-07-20). Lightning Ball (Ball Lightning) Created by Thunder, Shock-Wave. arXiv:1007.3348.

175. Coleman, PF (1993). An explanation for ball lightning?. Weather. 48: 30. Bibcode:1993Wthr...48...27.. doi:10.1002/j.1477-8696.1993.tb07221.x.

176. Meshcheryakov, Oleg (2007). Ball Lightning–Aerosol Electrochemical Power Source or A Cloud of Batteries(PDF). Nanoscale Res. Lett. 2 (3): 319–330. Bibcode:2007NRL.....2..319M. doi:10.1007/s11671-007-9068-2. Retrieved 27 June 2007.

177. Ball lightning›s frightening . . . but finally explained. EE Times. 29 August 2007.

178. Meshcheryakov, Oleg (1 August 2010). How and why electrostatic charge of combustible nanoparticles can radically change the mechanism and rate of their oxidation in humid atmosphere. arXiv:1008.0162 .

179. Meshcheryakov, Oleg (17 Jul 2013). Why not only electric discharge but even a minimum charge on the surface of highly sensitive explosives can catalyze their gradual exothermic decomposition and how a cloud of unipolar charged explosive particles turns into ball lightning. arXiv:1402.3214 .

180. Mario Rabinowitz (11 December 2002). Little Black Holes:Dark Matter And Ball Lightning. Astrophys. Space Sci. 262 (4): 391–410. arXiv:astro-ph/0212251 . Bibcode:1998Ap&SS.262..391R. doi:10.1023/A:1001865715833.

181. Muir, Hazel (23 December 2006). Blackholes in your backyard. New Scientist. 192 (2583/2584): 48–51. doi:10.1016/S0262-4079(06)61459-0.

182. http://www.disclosureproject.org/docs/pdf/uap_exec_summary_dec00.pdf | UAP in the UK Air Defence Region, Executive Summary, Page 7, Defence Intelligence Staff, 2000

183. Could some ball lightning observations be optical hallucinations caused by epileptic seizures, Cooray, G. and V. Cooray, The open access atmospheric science journal, vol. 2, pp. 101–105 (2008)

184. Peer, J.; Kendl, A. (2010). Transcranial stimulability of phosphenes by long lightning electromagnetic pulses. Physics Letters A. 374 (29): 2932–2935. arXiv:1005.1153 . Bibcode:2010PhLA..374.2932P. doi:10.1016/j.physleta.2010.05.023.

185. Erratum: Peer, J.; Cooray, V.; Cooray, G.; Kendl, A. (2010). Erratum and addendum to Transcranial stimulability of phosphenes by long lightning electromagnetic pulses. Physics Letters A. 347 (47): 4797–4799. Bibcode:2010PhLA..374.4797P. doi:10.1016/j.physleta.2010.09.071.

186. Ball lightning is all in the mind, say Austrian physicists, The Register, 19 May 2010.

187. Emerging Technology (2010-05-11). Magnetically Induced Hallucinations Explain Ball Lightning, Say Physicists – Technology Review. Technologyreview.com. Retrieved 21 January 2014.

188. Seward, C., Chen, C., Ware, K. Ball Lightning Explained as a Stable Plasma Toroid. PPPS- 2001 Pulsed Power Plasma Science Conference. June 2001.

189. Chen, C., Pakter, R., Seward, D. C. Equilibrium and Stability Properties of Self-Organized Electron Spiral Toroids. Physics of Plasmas. Vol. 8, No. 10, pp. 4441–4449. October 2001.

190. Seward, Clint. Ball Lightning Explanation Leading to Clean Energy. Amazon.com. 2011.

191. Manykin, E. A.; Ojovan, M. I.; Poluektov, P. P. (2006). Samartsev, Vitaly V, ed. Rydberg matter: Properties and decay. Proceedings of the SPIE. SPIE Proceedings. 6181: 618105. Bibcode:2006SPIE.6181E..05M. doi:10.1117/12.675004.

192. Norman, G. É. (2001). Rydberg matter as a metastable state of strongly nonideal plasma. Journal of Experimental and Theoretical Physics Letters. 73: 10–12. Bibcode:2001JETPL..73...10N. doi:10.1134/1.1355396.

193. Manykin, E. A.; Zelener, B. B.; Zelener, B. V. (2011). Thermodynamic and kinetic properties of nonideal Rydberg matter. JETP Letters. 92 (9): 630. Bibcode:2011JETPL..92..630M. doi:10.1134/S0021364010210125.

194. Holmlid, L. (2007). Direct observation of circular Rydberg electrons in a Rydberg matter surface layer by electronic circular dichroism. Journal of Physics: Condensed Matter. 19 (27): 276206–276202. Bibcode:2007JPCM...19A6206H. doi:10.1088/0953-8984/19/27/276206.

195. Endean, V. G. (1976). Ball lightning as electromagnetic energy. Nature. 263 (5580): 753–755. Bibcode:1976Natur.263..753E. doi:10.1038/263753a0.

196. Singer, Stanley (1971). The Nature of Ball Lightning. New York: Plenum Press.

197. The IEEE, Plasma Cosmology and Extreme Ball Lightning. Thunderbolts.info. Retrieved 13 July 2009.

198. Smirnov 1987, Physics Reports, (Review Section of Physical Letters), 152, No. 4, pp. 177–226.

199. Torchigin, V.P. (2009). Ball Lightning as an Optical Incoherent Space Spherical Soliton in Handbook of Solitons: Research, Technology and Applications editors S.P. Lang and S.H. Bedore. New York: Novapublishers: 3–54.

200. Wu, H.-C. (2016). Relativistic-microwave theory of ball lightning. Scientific Reports. Retrieved 22 June 2016.

CHAPTER 8

Microscale Meteorology

Chapter Contents

Meteorological analysis and prediction of microscale weather events like tornados, microburst, and other localized effects can be used as a predictive tool, or retrospectively to evaluate historical events. Applications include asset management, corporate risk management, strategic planning, building and engineering design assessment, and determination of agricultural impacts, to name a few.

INTRODUCTION

Microscale meteorology is the study of short-lived atmospheric phenomena smaller than mesoscale, about 1 km or less. These two branches of meteorology are sometimes grouped together as "mesoscale and microscale meteorology" (MMM) and together study all phenomena smaller than synoptic scale; that is they study features generally too small to be depicted on a weather map. These include small and generally fleeting cloud "puffs" and other small cloud features. Microscale meteorology controls the most important mixing and dilution processes in the atmosphere. Important topics in microscale meteorology include heat transfer and gas exchange between soil, vegetation, and/or surface water and the atmosphere caused by near-ground turbulence. Measuring these transport processes involves use of micrometeorological (or flux) towers. Variables often measured or derived include net radiation, sensible heat flux, latent heat flux, ground heat storage, and fluxes of trace gases important to the atmosphere, biosphere, and hydrosphere.

MESOSCALE METEOROLOGY

Mesoscale meteorology is the study of weather systems smaller than synoptic scale systems but larger than microscale and storm-scale cumulus systems. Horizontal dimensions generally range from around 5 kilometers to several hundred kilometers. Examples of mesoscale weather systems are sea breezes, squall lines, and mesoscale convective complexes.

Vertical velocity often equals or exceeds horizontal velocities in mesoscale meteorological systems due to nonhydrostatic processes such as buoyant acceleration of a rising thermal or acceleration through a narrow mountain pass.

A meso-beta scale vortex

Subclasses

Mesoscale Meteorology is divided into these subclasses:

- **Meso-alpha** 200–2000 km scale of phenomena like fronts, squall lines, mesoscale convective systems (MCS), tropical cyclones at the edge of synoptic scale
- **Meso-beta** 20–200 km scale of phenomena like sea breezes, lake effect snow storms

- **Meso-gamma** 2–20 km scale of phenomena like thunderstorm convection, complex terrain flows (at the edge to microscale, also known as storm-scale)

As a note, tropical and subtropical cyclones are classified by National Hurricane Center as synoptic scale rather than mesoscale.

Mesoscale Boundaries

As in synoptic frontal analysis, mesoscale analysis uses cold, warm, and occluded fronts on the mesoscale to help describe phenomena. On weather maps mesoscale fronts are depicted as smaller and with twice as many bumps or spikes as the synoptic variety. In the United States, opposition to the use of the mesoscale versions of fronts on weather analyses, has led to the use of an overarching symbol (a trough symbol) with a label of outflow boundary as the frontal notation.

REFERENCES

1. Orlanski, I. (1975). "A rational subdivision of scales for atmospheric processes". Bulletin of the American Meteorological Society. 56 (5): 527–530.
2. http://www.nhc.noaa.gov/aboutgloss.shtml
3. Roth, David. "Unified Surface Analysis Manual". Hydrometeorological Prediction Center. Retrieved 2006-10-24.
4. "AMS Glossary of Meteorology". Micrometeorology. American Meteorological Society. Retrieved 2008-04-12.
5. Rogers, R. (1989). A Short Course in Cloud Physics. Oxford: Butterworth-Heinemann. pp. 61–62. ISBN 0-7506-3215-1.
6. Pereira, Norman (1979). Air and Noise Pollution Control. Totowa: Humana Press. p. 200. ISBN 0-89603-001-6.

Index